On Line and On Paper

Inside Technology
edited by Wiebe E. Bijker, W. Bernard Carlson, and Trevor Pinch

On Line and On Paper

Visual Representations, Visual Culture, and Computer Graphics in Design Engineering

Kathryn Henderson

The MIT Press
Cambridge, Massachusetts
London, England

This book was set in Baskerville by Achorn Graphic Services Inc. and was printed and bound in the United States of America.

Library of Congress Cataloging-in-Publication Data

Henderson, Kathryn.
  On line and on paper: visual representations, visual culture, and computer
  graphics in design engineering / Kathryn Henderson.
    p.   cm.—(Inside technology)
  Includes bibliographical references and index.
  ISBN 0-262-08269-1 (hc: alk. paper)
  1. Engineering design. 2. Computer graphics. 3. Visual communication.
I. Title. II. Series.
TA174.H458   1998
620'.0042'0285—dc21   98-5211
                                                                    CIP

Contents

Preface
Messy Questions Pursuing Messy Practice

Research questions, like life and work, are not static but messy. Since they are connected to the experiences and networks of the researcher, they change as the investigation evolves. So, too, do questioners change with their evolving questions. The questions that launched this research began to take form while I was testing my wings as an art critic. My years spent learning to sketch and draw made it clear to me that artistic rendering skills were not enrichment frills that were so trivial that school boards could cut them with impunity. I suspected, rather, that they were connected to cognitive skills as basic as mathematics and verbal literacy and equally applicable to all sorts of problem solving—in math and science as well as the visual arts. Learning to draw improved my capacity to see things precisely, in fine-grained detail, but it also influenced the way I process and manipulate visual information. The more I look at the formal components of an artwork—structure, brush stroke, texture, palette, depth or flatness of space, narrative or abstract content—the more I recall other contextual details. Historical, biographical, and contextual information about the artist and the period in which she or he worked is indexed by the visual representation and the process of analyzing its formal characteristics. Although my drawing skills have deteriorated without continued practice, the cognitive abilities have remained.

I began to ask scientists, engineers, and mathematicians if they had similar cross-discipline experiences. In one college class, the professor coached students to visualize the double helix. In some textbooks, math problems and brain teasers required visualizations such as up-the-mountain, down-the-mountain scenarios for graphing and up-the-river, down-the-river scenarios for setting up algebraic equations. I was reminded of the visual correlations between certain equations and their graphed shapes—parabola, ellipse, circle—and recalled my own junior high biology assignments to draw what I saw in the microscope. And yet a link between visual skills and math and science skills does not seem to be recognized at the level of public school curricula, the focus for most popular concern about declining achievement in math and science. My next question, then, was, Why is the importance of visual skills as cognitive skills that are applicable in math and science not acknowledged? Why are visual skills taken for granted and hence haphazardly taught?

I found several books helpful as I thought about these issues. Patricia Greenfield's (1984) account of how children develop visual literacy documented visual knowledge in practice. Investigating her citations, I found Eugene Ferguson's (1977) account of how technical knowledge was transferred almost solely through visual representations from the 1400s to the recent past. Viewing this as evidence of the former high status of visual knowledge, my question changed to, Why does visual knowledge have such low status today in science and engineering if it enjoyed high status in the past?

I soon realized that my assumption of any consistent status for visual knowledge was problematic. A debate about the reliability of the seen is ongoing. It was first recorded by the Greeks and continues through contemporary challenges to ethnographic observation and concern over the proliferation of electronic imaging in computers and video, which is sometimes seen as a threat to verbal literacy. The status issue crystallized when I interviewed the engineering design professor who taught the first practical computer graphics course offered in my university's theory-oriented engineering school. Not only did he have to hire a local junior college instructor to teach the drafting portion of the course because drafting was not regarded as appropriate curriculum at the university, but his students had trouble mastering visualization skills (such as turning an object 180 degrees in their mind's eye) because they regarded it as a trivial accomplishment that was unworthy of their efforts. My question changed to, What historical and contemporary factors have contributed to the present status of visual representation in science and technology?

I spent hours in the library, examining the period when science became separated from art and finding Alberti's fascinating argument to the Medici court in the 1400s that the rationalization of depicted space, accomplished through the techniques of perspective, rendered painting a science and Zilsel's account of the codification of tacit and rule-of-thumb shop knowledge as the beginning of scientific theorizing. Given that language and mathematics are highly structured and codified, I began to think that codification was the key to the status question and pursued appropriate literature. But I wanted to get out of the library and into the world of work to answer my next question: What is going on today with visual representations in engineering design practice?

Using contacts from the university alumni directory, I visited my first field site and confirmed my suspicions that design engineers still communicate in a multitude of visual formats, the newest being computer graphics. This brought up new questions. I did comparative interviews in lunchrooms and lobbies and on sidewalks and patios at engineering sites to ensure that my initial findings were not an isolated instance. Many firms were using computer graphics only in their promotional literature and not in design, but no one in town would let me into their inner sanctum of design for a second case study. I once was even accused of being an industrial spy! Finally, through contacts made at a workshop run by a computer graphics consultant, I met a graphics manager from a city two hours away from the university where I finally was given access for participant observation. From this new setting, new questions arose. Finally, I went to the San Francisco Bay area to interview workshop contacts in civil, electronic, and mechanical engineering.

The questions that evolved along with the research, like engineering design and technology construction, did not follow a clean linear process. Rather, they peeked out from behind bushes and jumped up hither and yon, forcing me to chase down answers in a messy manner—similar to one design engineer's description of his workday as "putting out one fire after the other." My hope is that the answers to those squirmy, messy questions will convey that other ways of knowing are important and that visual documents and the interactions surrounding their construction and use are crucial aspects of technoscientific practice.

This book offers tentative answers to questions that grew out of my experiences, my observations, and my assumption that the most powerful knowledge is based in concrete practice. Just as workers' knowledge is based in their practice, so too is the information I am able to share here based in my practice, in all of its messiness. Part of that practice included mentoring from my academic family. I wish to thank Eugene Ferguson for his thoughtful comments and many letters; Bruno Latour, Karin Knorr Cetina, Leigh Star, and Judith Perrolle for feedback and encouragement; Peter Meiksins for his generosity with engineering history bibliographies; and Peter Whalley, Ron Westrum, the editors of the Inside Technology series at the MIT Press, and the anonymous reviewers for comments on earlier drafts. My gratitude to my academic uncles: Bud Mehan, Michael Schudson, Robert Horowitz, and Joe Gusfield for guidance and for reminding me that my questions were a lifetime's work of which I could address only a small piece at a time. To Mary Ellen Boyle, thank you for pushing me to send out the manuscript; to Mary Zey, Renee Rabb, and Deborah Black, thank you for help with the messy final details. Most especially thank you to Chandra Mukerji for her always insightful and pragmatic advice on everything one needs to know to write a first book. Any missteps are, of course, my own and no reflection on these good people.

To all the engineers, designers, drafters, and managers who put up with my intrusion into their work, my sincere gratitude for taking the time to answer my questions. Thanks also to the consultant who allowed me to attend his workshop, which subsequently allowed me access to such excellent informants, and my appreciation to the companies that must remain anonymous here but granted me permission to observe at their sites and to reproduce some of their documents.

And to my three daughters, Patricia, Christina, and Theresa, thank you for being there to ground me in daily life in those moments when I had doubts about being a scholar and thank you, also, to the University of California, San Diego, community for supporting my alternative identity during my daughters' messy passage through adolescence, when I had doubts about being a mother. To my own mother, Henrietta Henderson, now deceased, gratitude for unflagging moral and financial support all her days. To her I dedicate this book.

On Line and On Paper

1

Introduction

A senior drafter, recently promoted to engineer, fights with management to get her drawing board back, arguing, "I can't think without my drawing board." A field engineer, explaining the workings of a turbine engine to a group of third-world users who speak a variety of languages, says, "Thank God for the drawings": he had used colored pencil to differentiate functions. A chip designer at a computer firm keeps a stack of paper printouts near his computer screen to help him keep a new chip's overall design in mind. Despite the array of electronic tools as his disposal, he returns to paper to analyze his designs and correct errors.

In the world of engineers and designers, sketches and drawings are the basic components of communication; words are built around them. Visual representations are so central that people assembled in meetings wait while individuals fetch drawings from their office or sketch facsimiles on whiteboards. Coordination and conflict take place over, on, and through drawings. Visual representations shape the structure of the work and determine who participates in that work and what its final products will be. They are a central component of a social organization based on collective ways of knowing. This culture of visual communication has been strongly affected by computer graphics, and a primary goal of this book is to understand the changes triggered by innovation.

To understand the visual culture and technical work of engineering design, we must observe the daily processes involved in the work itself. In sociology this sort of ethnographic approach is usually grounded in some form of participant observation. This means that the researcher participates in the everyday activities of the group she is studying and makes her best effort to understand that world from the perspective of the insiders, in the same way that anthropologists approach the ethnography of cultures different from their own. To be a participant observer, the researcher must learn everything necessary to "pass as a native"

(Goodenough 1957, 1971), including but not limited to displaying the appropriate behaviors and demeanors in a variety of circumstances, understanding which actions are regarded as important or trivial under varying conditions, and exhibiting competency in communication using local terminology in the appropriate way at the appropriate time. No matter how exotic or commonplace a culture might appear, *every* group has special characteristics and idiosyncratic behaviors that an ethnographer must respect.

The point, as Davis (1973) has noted, is to maintain the delicate balance between the perspective of the convert and that of the Martian. If a researcher becomes too entrenched within a culture, everything appears as natural and given—"just the way things are." But if she stands too far outside a culture, without a grounding in members' meanings, she may miss important elements of context and therefore misunderstand events. Participant observation drenches the researcher in context but also allows her to step back for reflection and analysis. Reflexivity—the act of constantly examining and questioning everything about the research process, including the researcher's role and impact on the research setting and the contextual elements influencing members' accounts—has long been part of the sociological tradition. As the research proceeds, the investigator constantly checks perceptions with those of the informants and gathers additional data through formal individual and group interviews and ordinary conversations.

Theory, too, is grounded in observations and then checked and refined against further observations as the research proceeds. Using such a grounded theory approach (Glaser and Strauss 1967, Strauss and Corbin 1994) does not, however, preclude entering a dialogue with other theories and theorists and building on their work. In the present study, the dialogue is with the growing body of research in science and technology studies.

The study of science and technology is not new to sociology, but some significant new perspectives have arisen in recent years. Prior to the late 1970s the dominant paradigm, established in American sociology by Robert Merton (1942, 1973), focused on the norms of science, its reward system, and its career structure—what one might call the *sociology of scientists* rather than the sociology of science or scientific knowledge. A new generation of scholars, many European, began in the 1970s and 1980s to extend sociological analysis to the content of science. While these authors differ in their methodological and theoretical approaches, their common ground includes a curiosity about what scientists actually do in their workplaces and a challenge to the persistent mythology that attributes technological accomplishments to scientific truth applied to human needs. Also prominent in this perspective is an often misunderstood rejection of the distinction between good and bad science and between successful and unsuccessful scientific theory. Rather than focusing on the truth or falsity of a scientific theory, these researchers concern themselves with the process by which the status of truth or falsity is established.

This line of research was originally influenced by Thomas Kuhn's *The Structure of Scientific Revolutions* (1962), which raised intriguing questions about the cumulative nature of scientific knowledge and the circumstances in which paradigm shifts occur. A number of researchers who explored the consequences of this view were at the University of Edinburgh, and the *interests model*—which argues that science is not neutral but that the outcome and content of science as well as access to it as a profession are determined by structural commitments, political positions, and other institutional considerations—became especially associated with the Edinburgh school (see, for instance, Barnes 1977). Comparable antipositivist approaches were pursued in Paris, in Amsterdam, and in San Francisco at the Tremont Research Institute. A common element of all these programs was an interest in opening up the black boxes of received scientific fact and technological artifact to analyze the process of their production and thereby demystify them. Often the research methods were similar to the ethnographic approach described above, though the antipositivist researchers derived their methodologies from a variety of philosophical, historical, anthropological, and other traditions in addition to those of sociology. Latour and Woolgar's *Laboratory Life: The Social Construction of Scientific Facts* (1979), probably the best-known ethnographic study of a scientific laboratory, documents the creation of a scientific fact by following technoscientists around.

As in science, investigation of the social construction of technology necessarily leads to the examination of daily working life. Scholars in science and technology studies (sometimes also known as the *sociology of scientific knowledge*) view the science-technology relationship as egalitarian and interactive. To get closer to the technical content of discovery, experimentation, replication, argumentation, and representation as they occur in the processes of knowledge production, a growing number of researchers have incorporated constructivist and ethnomethodological approaches into their work, including participant observation, in-depth interviews, and analysis of informal shop talk (see, for example,

Latour and Woolgar 1979; Knorr Cetina 1981; Lynch 1985a; Collins 1985).

Another important perspective in science and technology studies is actor-network theory, developed by Michel Callon (1986), Bruno Latour (1987), and John Law (1987b). Its central argument is that successful technological innovations are achieved through the construction of durable and interconnected links that tie humans and nonhuman entities together as mutual actors (*actants*). Law calls this *heterogeneous engineering*. As an example, in chapter 5 it is shown that moving a new surgical instrument design into production takes technicians, engineers, doctors, patients, cataracts, rabbits, lawyers, biologists, clean rooms, marketing agents, computer-generated and hand-drawn drawings, infusion plastic, molds, an eager molding company, an antagonistic manufacturing department, and forceful management along with the political meanderings of prototypes.

One important contribution of the actor-network perspective is its emphasis on the mutually reinforcing relationship between a technology and its network: as one changes, so does the other, in both directions. The social shaping of technology is not merely a matter of static social networks shaping malleable technology: heterogeneous engineering affects social relationships as well. As Winner (1980) puts it, "Artifacts have politics." Memorable among his several examples are the Long Island bridges of master contractor Robert Moses, who intentionally built them too low for buses that might have carried inner-city riders to a beach area Moses helped construct and wanted to reserve for those who could afford automobiles.

Another important contribution of actor-network theory is its symmetrical treatment of nonhuman and human players in the network. Such inclusion reminds us that technology and society are not independent but are symbiotically linked. While this idea has gained fairly wide acceptance in technology studies, the idea of symmetry has raised some controversy. The most significant challenge comes from Collins and Yearly (1992), who maintain that the symmetrical analysis of human and nonhuman actants forces us to forego the symmetry between the truth and falsity of scientific belief, since we are dependent on human accounts of the behavior of nonhumans: that is, since the normatively accepted account of nonhuman behavior is a scientific one, scientific truth becomes privileged. This debate in some ways reintroduces the older issue of the place of the real or material world in explanations provided by the sociology of knowledge. It was, in fact, in order to reframe this realist versus relativist debate that Latour (1987, 1988a, 1992) and Callon (1986) included nonhumans in the network of actants involved in the social construction of a scientific fact or technological artifact.

Donald MacKenzie (1996) rightly points out that contending with materiality does not have to be an either/or proposition but can be understood in terms of multiple accounts and transitions. In his example of the impassioned debate between the Patriot missile system's defenders and critics in the aftermath of the Gulf War, the two camps drew radically opposed conclusions from events that took place above the territories of Israel and Saudi Arabia. MacKenzie sees the crucial point here as the distinction between what Barnes and Bloor (1982) have termed *unverbalized reality* and our beliefs and descriptions about a given reality. The crucial moment, then, occurs when unarticulated reality (which actor-network theory includes as the action of nonhuman actants) is transformed into verbal accounts of that action, where Collins and Yearly's critique begins to apply.

Multiple readings of material reality and the transition from the action of nonhuman actants to human accounts of that action are themes interwoven throughout this study. In chapter 4 multiple readings of drawings and prototypes from the perspectives of different engineering specialties, corporate accounting, marketing, and inventory managers are shown to facilitate coordination in the design-to-production transition of a turbine engine package. This capacity of artifacts to be read on different levels by different groups has been addressed in the sociology of culture at least since Herbert Gans (1974) examined the differing interpretations of literature that flow from the class background of the reader. Leigh Star first raised the issue in science and technology studies by introducing the term *boundary object* for material objects that facilitate the coordination of scientific work because they can be interpreted in a tightly focused way by specialists while being simultaneously readable by generalists (Star 1989; Star and Griesemer 1989). While this concept allows us to see coordination, as in the case of the Patriot missile system, multiple readings of the action of actants can also be in conflict. In chapter 5 the political conflict between a research and development group and a manufacturing department contributes to conflicting readings of how a medical instrument prototype performs at an ophthalmological exhibition. The crucial moment of transition, when the dispute is settled, occurs when the prototype is circulated within a network of highly placed members whose status helps resolve the controversy in favor of the design.

The data presented here—based on my participation in the everyday activities of designers and drafters, technicians and managers—illustrate the social and practical consequences of replacing interactive social practices with technical equipment. Normal interaction in these situations depends on creating a social organization of material objects and tacit knowledge through visual conventions and constructing and negotiating the collectively shared knowledge intrinsic to team design work.

Tacit knowledge—knowledge that is not verbalized, sometimes because it is taken for granted but often because it is not verbalizable—has become another focal point as STS (science and technology studies) researchers have questioned how scientists and engineers know what they know. Polanyi (1958, 1967) introduced the term to explain *experiential knowledge,* such as a carpenter's knowledge of how to choose the appropriate nail for a particular kind of wood or the way humans normally recognize a face. It is viewed here as a residual category, a holding place for many unverbalized forms of knowing, including but not limited to knowledge gained through visual elements and touch.

Technological innovation emerges out of the mundane interactions of actors, machines, and paper. The work presented here joins that of other STS researchers in challenging the view that technology is driven by science.[1] Engineering sketches and drawings are the building blocks of technological design and production. Moreover, because they are developed and used in interactions, these visual representations act as the means for organizing the design-to-production process and hence serve as a social glue both between individuals and between groups. The drawings and sketches themselves structure the work process as well as its product.

Representational practice in science has been increasingly a focus of examinations of the production of scientific knowledge. Studies in this area pay close attention to what Latour and Woolgar (1979) have termed *inscriptions*—visual or verbal devices such as intentionally arranged, trimmed, and filtered samples, carefully cropped photographic records, and edited chart tracings (Amann and Knorr Cetina 1988; Latour 1986; Lynch 1985b, 1988; Pinch 1985; Rudwick 1976; Shapin 1984).[2] The view of technological practice taken in this study similarly shows technology as a social process, simultaneously socially shaped and shaping society. Studies of the social construction of technology by Ackrich (1992), Ackrich and Latour (1992), Bowker (1987, 1994), Law (1987a, 1987b), Law and Callon (1988), MacKenzie (1990), and MacKenzie and Spinardi (1988), to name a few, illustrate the view that the received constructed environment is not the only one possible and that the processes of drawing, writing, recording, inscribing, constructing, and persuading over time must be understood as part of the production of any technological artifact. I join these scholars in the conviction that all knowledge and knowledge claims are socially constructed and in arguing for a nonreductionist, interactive approach to the machine in context.

While a good number of studies taking such an approach have looked at representational practice using historical materials (Rudwick 1976, 1992; Ferguson 1977, 1992; Mukerji 1984; Bowker 1988; Gooding 1990, 1993; Gorman and Carlson 1990, 1993), the majority of such studies of contemporary representational practice have concentrated on the physical and biological sciences (Lynch 1985b, 1988; Yoxen 1987; Amann and Knorr Cetina 1988; Meyers 1988; Knorr Cetina 1990; Ruse and Taylor 1991; Pinch and Collins 1993). Engineering design studies have noticed the importance of visual documents in the network of people and things that play a role in technology design (Bucciarelli 1994, Vincenti 1990) but have not focused on it.

This book looks specifically at the role of visual representations in everyday engineering practice. It is based on participant observation of practices in two contemporary industrial settings among engineers engaged in the process of designing new pieces of technology.[3] For over a year I worked as a technical writer at an engineering firm that produces turbine engines. I attended design meetings, interviewed design engineers, and observed the interaction of designers and shop workers as they built a new prototype engine package. Subsequently, at a firm that produces high-precision medical optics, I observed informal R&D working sessions, laboratory test procedures, and more formal design meetings during the development of a new surgical instrument. Finally, to serve as a check on these data, I conducted in-depth, open-ended interviews with engineering designers and managers in approximately thirty other West Coast firms. Throughout quoted materials transcripts have been only minimally cleaned and edited in order to retain individual voices and a grounding in actual spoken language. All individual, company, and product names are pseudonyms.

My conclusion from these studies, unsurprisingly, is that there is no one best way to use a computer-graphics system; firms and individuals engage in many types of mixed paper and electronic practices as well as differential uses of electronic options. In many companies, experienced design engineers work by hand, and drafters then redraw the designs onto graphics systems. Many designers switch between hand drawings, hard copies from the computer, and the monitor screen, often preferring to work out initial concepts in hand-drawn sketches and returning to paper copies during analytical phases of their work.

What I have seen in these cases is that the computer–driven implementation of a single right way to render graphic representations can destroy practices that are important for specific aspects of design activity. The designers of many computer-assisted design systems assume that innovation is a rigidly linear process, going from idea to drawing to prototype to production. My research suggests—as Latour and Knorr Cetina, among others, have argued—that such rigid models are seriously misguided. While such models integrate well with and are even prerequisite to the construction of an integrated computer system, the assumptions on which they are based and the very efficiency of their intended use break down social communication practices that normally repair the frequently occurring problems and misunderstandings that are part of the work process in a world of messy practice.

The vital necessity of visual information in the technological practices of engineers from the fifteenth century forward has been well documented by historians. In the earliest examples visual representations were so crucial to technological endeavors that drawings (which, of course, indexed tacit knowledge) were the entire visible carriers of any encoded building information, with little written or mathematical elaboration (Booker 1963; Ferguson 1977, 1992; Baynes and Pugh 1981; Hindle 1981). The more recent application of computer-generated graphics to engineering design work has highlighted the contrast between design engineers' actual practice and engineering school protocols (Ullman, Stauffer, and Dietterich 1987). Some authors have suggested that the process of creating drawings may be as important to design as the drawings themselves (Bly 1988; Tang and Leifer 1988; Bucciarelli 1994). My examination of visual communication practices in the world of technological innovation focuses on visual documents and the interactions around them that together structure the design process and hence reveal changes in the structure of interaction among actors and artifacts when computer-aided design is brought into the process.

The questions that drove this research are built on an assumption that the knowledge used in everyday work is grounded in practice, the learning of practice, the history of a given practice, and the cultural, technical, and organizational constraints constructed around practice (Hutchins 1995; Lave 1988; Suchman 1987). Before we can understand what kind of changes computer graphics are bringing to engineering design practice, we must understand the base on which current practice is built. For this reason chapters 2 and 3 address the history of engineering design practice as linked to the history of engineering. Chapter 2 asks, What role has the codification of knowledge played in the historical development of professions in general and engineering in particular? Chapter 3 asks, What are the roots of the visual culture of engineering? Is it a kind of codification of knowledge? Where did the conventions of rendering mechanical things come from? How do these conventions influence the way engineers see their world? The answers to these questions reveal an engineering visual culture—a particular way of seeing the world that is explicitly linked to actual material experience in rendering that world, defining in Latour's words (1986, 12) both "what it is to see and what there is to see." This chapter also develops the concept of visual representations as conscription devices to explain the intersection between the roles of inscriptions and boundary objects, which facilitate distributed cognition in team design work.

The next two chapters move from historical perspectives to contemporary ethnography. Chapter 4, the first case study, addresses two basic questions: How are visual representations being used in industrial design and production today? And are the visual practices of engineers changing because of the introduction of computerized graphics systems? This study follows what one engineer aptly titled "the yellow brick road"—the path from design to production of a turbine engine package. It documents the variety of paper-world visual representations employed along the way as the members of the design team, like Dorothy and her buddies, try to overcome various obstacles that could deter them from their goal, such as a hostile environment due to downsizing, a labor strike, misinterpretations of drawings, and the problems of dealing with a new computer-graphics technology that could allow a novice user to lose a whole day's work in one stroke. The innovative methods that workers develop to work around obstacles, known as "workarounds" are traced as they trek onward to produce their prototype.

Chapter 5, the second case study, answers several levels of questions, starting with, How do prototypes fit into the cascade of sketches from initial idea to final iteration? The transformation of the design of a highly innovative surgical tool is followed as the social organization of participants and participation change with subsequent visual and actual renditions of an emerging prototype. Conflict involving the design and the visual documents was an issue at this site. While the turbine engine network in the first case study had to contend with conflict during the production of its prototype, due to labor unrest, this did not directly affect the acceptance of the design. The research and development team in the second case study had to contend with interdepartment conflict challenging the adequacy of the design, which offered an opportunity to answer the question, What role do visual representations and computer graphics play when conflict occurs within the design-to-production network? The analysis reveals that visual representations, including prototypes, are not only devices for communal sharing of ideas but are also a ground for design conflict and company politics, exactly because they facilitate the social organization of workers, the work process, and the concepts that workers manipulate to produce a collective product. In this charged climate, computer-aided design was marginalized to serve merely as a record-keeping tool for drafters rather than as a design tool for engineers. This allowed the design team to preserve its visual culture and support network intact against the potential for a reorganization that could change political and power alignments.

Chapters 6 and 7 report the results of a series of interviews I conducted at a variety of engineering firms. Chapter 6 asks, What kinds of restructuring does the implementation of a new graphics system bring about? It looks at the proliferation of interlocking restructurings, including (1) changes in the official job status and responsibilities of those who produce design work, (2) changes that affect the work itself, at the level of the individual interacting with the design, and (3) changes that affect the structure of work at the group level, where people and companies interact with one another as well as the design. Chapter 7 addresses the consequent question, How do people who use the new technology cope with the restructuring it causes? This chapter looks at the practices that design engineers, drafters, and managers employ to get the work done despite the disruptive restructurings computer-assisted design has engendered.

Despite the many workarounds and mixed practices needed to make graphics systems work, the positivistic language of computer culture permeated my fieldwork conversations. Throughout my observations the term *high tech* occurred regularly in discussions of computer graphics. To gain a better sense of the meanings ascribed to this phrase, I began to query engineers for their definition of *high technology*. Chapter 8 surveys their answers and the further questions those answers spawned. Though the field question was simple—How do practicing engineers define the term *high tech*?—the answers pointed to much deeper issues.

Many practicing engineers were unwilling to give glib definitions. Those who gave it thoughtful consideration seemed uncomfortable with the term, and some saw it as more like marketing language than engineering language. Analysis of their definitions provides not only a cultural ethnography of members' diverse meanings for a term, but a new set of revealing questions that address why technical workers would put up with the drastic rearrangement of their visual culture that computer graphics engenders. These new questions include, What is the role of standardization in engineering? How do new technologies invoke mystification and status? Do interesting relationships exist among concepts such as high art, high culture, and high technology?

In answering these questions, the chapter opens access to the wider context of engineers' mixed practices. That design engineers' innovative practices are mixed is important. In contemporary and historical times engineers have not been resistant to new technologies or to standardization but rather have filled in the gaps to make them work. Indeed, as chapter 2 points out, the historical professionalization of engineering has been built on the standardization of parts, procedures, and people. However, engineering design work is composed of both art and standardized codings, and both must be protected. Mixed practices are actions taken to protect the cognitive spaces where visual creativity takes place. But engineers must also protect engineering as a profession even as they concentrate on doing their daily work. To understand this bifurcation of allegiance, or perhaps nesting of interests, this chapter discusses the never-ending quest for the constantly elusive, latest and greatest technology—the aura of high tech—which loses its glamour as soon as it is widely understood. To understand such an addiction to the latest and greatest technology, the chapter revisits the message of chapter 2, the professionalization of engineering through the codification and standardization of both machines and human knowledge. Reviewing engineering's historical professionalization project as a series of coding forays into the knowledge base of the machine shop reveals the strength of such standardization. The main accomplishment of such codification is in claiming a knowledge domain for commercial advantage, while another, tacit portion of knowledge remains uncodified. It is this unarticulated knowledge, often visual in format, that is the necessary basis for creative design work. It is also this knowledge that can be most recalcitrant to the use of graphics systems; and for this reason the conflict between using codified and uncodified knowledge necessitates the use of mixed electronic and paper practices.

In concluding, I take the issue of codification a step further, addressing the question, Why is visual representation so powerful? I develop here the concept of the meta-indexical role of visual representations, which allows them to be more than the sum of their parts. They serve as a holding ground where codified and uncodified knowledge can meet, drawing on each other's strengths to bring together various levels of tacit knowledge, including but not limited to visual knowledge, kinesthetic knowledge, mathematical knowledge (verbal and nonverbal), local and experiential knowledge, and multivisual competencies. This is why they are amenable to mixed-use practices. Such practices take advantage of the strengths of codified and uncodified knowledge as well as the in-between areas that cannot and should not be so simply dichotomized. It is this chameleon, meta–indexical, elastic quality of visual representations that demands mixed practices in the workplace and facilitates multivisual competency to enhance the creativity of individual and group design work.

Because visual practices are so important to problem solving, design engineers find ways to shield these practices from the formalizing structures of computer graphics systems or to mediate them through mixed–use practices and hence create a split between company–dictated procedures and informal work practices. My results suggest an intersection of concerns among the new science studies that are directed to the messy, visually oriented processes in the laboratories of technoscientists, labor studies of the deskilling effects of automation,[4] and ethnographic research on the importance of paper documents in collaborative work.[5] My focus on the way computer-graphics systems are simultaneously incorporated, resisted, and mediated shows the importance of day-to-day visual practices that make the application of fully integrated formal and rational systems problematic.

Even if systems do not run smoothly and it takes mixed practices to get the job done, however, industry representatives will insist that computer graphics and official documentation have accomplished the task rationally and efficiently. Marketing departments distribute illustrations with a high-tech look, such as the sculpted gridlike web of wire model drawings, to customers and top management—but these illustrations often are made after the design work has been completed. This practice is not unlike the historical use of perspective in detailed renderings of ships and machines in order to attract investors and the use of less dramatic plan, profile, and section drawings as the actual workhorses of production (Baynes and Pugh 1981).

I want to emphasize that this book does not make an argument against using computer tools but attempts to lay the groundwork for creating more usable tools. There are no one-way relationships among machines, people, mental models, representations, and constructed technology. Instead, messy, interactive practices contribute to the mutual and simultaneous construction of technological outcomes and the conscription network that produces them. Visual representations are both the product of and resources for situated practice in the heterogeneous elements that are part of engineering new technologies. This study is a step toward understanding such interactive social mechanisms associated with the visual conventions of engineering. With more research focused on the situated practice of engineering design work, we should be able to construct better-informed, more useful computer tools. As a first step to understand the links between practice and the larger interests of engineering, however, it is necessary to look backward rather than forward to the historical context of the codification of engineering knowledge—the subject of chapter 2.

2

Coding and Claiming: Codification and the Rise of Engineering as a Profession

In many traditional societies, one elementary form of political power is the quasi-magical power to name and thereby to make exist—that is, naming is thought to imbue the maker with power over the named (Bourdieu 1985). The counterpart in modern Western society, as Max Weber first noted, is the claim of legitimate domination based on rationalization (Gerth and Mills 1946). Basing actions on goal-oriented reason leads to organizing and hence codifying and standardizing which—like the act of naming—makes apparent a realm of knowledge and thereby makes it exist in the social world. Codifying legitimates the codifying agents' claims of power over a realm of knowledge. This chapter sketches the role played by codification and standardization in the emergence of engineering as a profession in the United States.

Given the historical evidence that visual communication and visual knowledge have always been crucial elements of engineering design (Ferguson 1977, 1992; Booker 1963; Hindle 1981), I can phrase my underlying question as follows: What is the relationship between the prestige status of engineers' visual skills and knowledge and the codification of that knowledge base? To answer this question I present the history of the professionalization of engineering as a series of attempts to codify the nonverbal knowledge base of master craftsmen[1] and machine shop culture.[2] This allows me to take some initial steps in linking engineers' actual work practices with the role and power of engineering as a profession.

Codified Knowledge as a Basis for the Professions

Standardizing, codifying, and ordering—ways of organizing knowledge into an ordered set—are part of any professionalization process. By *codification* I mean the setting of a body of knowledge into an ordered format based on a consistent strategy and some existing or newly created standardized symbol system.

Though codification may appear to be a neutral action, it often has been used as a rationale for excluding some people from access to a knowledge base. Freidson (1986) points out that those who develop higher knowledge usually address elite groups who share some of that knowledge and a belief in its virtues, but not laypeople or tradespeople who might have a more basic everyday experience of that knowledge. This process creates a distinction between everyday or commonsense knowledge, shared by all normal adults, and specialized knowledge, shared by particular groups who regularly perform now professionalized tasks.

In ancient civilizations as well as our own, people have differentiated between knowledge domains—sacred versus profane, theoretical versus practical, elite versus popular. In the West we have labeled knowledge as higher if it is theoretical and abstract and has been systematized in a way that allows exploration of the fundamental processes that constitute the world. The formalization of such knowledge, which I call *primary codification,* marks this knowledge as distinct and helps keep it separate from both common knowledge and nonformal specialized knowledge, while simultaneously restricting access to it.

Larson (1977) has argued that the demarcation of professions is a historic product based specifically on a market project—an attempt to translate specialized knowledge and skills into social and economic rewards. The act of demarcation is accomplished through a system of education, credentialing, and professional association. Larson's institutional level is a *secondary level of codification* that can occur only following the primary codification of knowledge into an ordered symbolic form of some sort. While for many professions the format is in the verbal domain, for engineering it is a combination of the verbal, the visual, the kinesthetic, and the mathematical.

This primary act of codification is crucial because uncodified knowledge is in some sense invisible and therefore cannot generate and facilitate economic monopoly rights for holders and users of a knowledge base, despite the fact that such tacit knowledge is the basis of everyday practice. The more highly codified a knowledge base is, the more visibility it has, and thus the more easily it can be used to impart status to the group that claims it.

How knowledge is structured so as to be acknowledged as being worthy of attention is particularly significant. While it is necessary to look at the characteristics and interests of both the carriers or agents of formal knowledge and the institutions that make their activities as agents of knowledge viable, it is equally important to look at how a reputation of high status and a market monopoly initially are created for a body of knowledge. This transition period for the engineering profession was when engineering schools in the United States first began to make a claim on machine shop knowledge.

Standardization and Codification in Engineering History

Standardization and codification have long been part of the knowledge claims of American engineering. In the nineteenth century the standardization of engineering training in technical schools opened the profession to middle-class aspirants and allowed their graduates to claim a monopoly on engineering knowledge. This claim cut off promotions from the shop floor (Calvert 1967), but the history of the two groups—school- and shop-trained engineers—cannot be seen simply as one of class conflict. In the last two decades of the century, both groups were interested in systematizing and standardizing the materials with which they worked, the methods they used for measurements, and the tools they used to work their materials into articulated machines. That standardization could be advantageous to almost all interests is reasonable.[3] The question was whose standards were to be adopted.

The American Society of Mechanical Engineers (ASME), for example, formulated a uniform code of methods for testing metals that allowed the results of different experimenters to be compared. This code reflected a common interest in rationalizing the nomenclature of the machine and the shop. As Calvert (1967) points out, however, this standardization was mainly the product of a generation of engineering textbooks and handbooks. The subsequent consistent use of this specialized nomenclature created de facto standards that worked to the advantage of people who had been trained with those texts.

While attempts were made to standardize almost every aspect of the machine shop, the question of whether standards would be set by an agency or arbitrated in the marketplace split the shop-trained entrepreneurs from the representatives of the technical schools. From 1888 to 1893, for example, there was a fierce battle within the ASME over whether the metric system should be adopted and recommended as a U.S. national standard. The shop people opposed conversion because it would cost them money without leading to any appreciable gain in efficiency. Educators supported conversion because it would add one more piece of curricular knowledge that could distinguish their graduates. Moreover, as Calvert (1967) points out, metric conversion would allow relatively complicated calculations to be done in seconds, thus making mental calculation a simple matter for the technical graduate. The conversion issue is interesting because it was one of the few arguments that the shop culture won. In engineering, as in other disciplines, the acquisition of theoretical bases and the methodical systematization of empirical knowledge eventually defeated apprenticeship, and educators took the lead in efforts at professionalization.

Technical content, of course, is only a part of practice. The next stage of codification involved work processes. Members of the shop culture elite supported codification of those processes on the assumption that this would increase productivity and thereby benefit their profits as well as their employees' wages. The turn of the century thus found the ASME debating the merits of scientific management.

Taylorism and the Rage to Codify, Classify, and Bureaucratize

Calvert (1967) is puzzled by the contradictory opinions regarding management systems held by men like Frederick Taylor, Henry Towne, and Frederick Halsey, who had come up through the shop culture system to become part of its elite and were active in the Progressive Era (1890–1920). He points out that even as these men promulgated systems of management that could have no other effect than to destroy shop culture as they had known it, they gave passionate lip service to the ideal represented by that culture. He finds Taylor a most curious figure, the giant of the scientific management movement who mercilessly attacked the personal, entrepreneurial, and unsystematic management of the machine shops while at the same time enthusiastically defending the shop precisely for allowing learning based on deep, personal experiences—"the best education and experience a man could get."

I find this not at all curious. Taylor, obsessed with the truth of science but experienced in the stimulating use of tacit knowledge and multiple ways of knowing involved in shop innovation and production, attempted to isolate and codify just one aspect of shop work—kinesthetic or manual knowledge. His intention was to create a technical elite—an altruistic corps with a service ethic—that could stand between labor and capital, advising both of the "one best means" of production, which would be discovered through the scientific method. Management, however, adopted a reified and standardized form of the tacit manual knowledge Taylor had taken from the shop floor, through time and motion studies, to regulate *all* aspects of work. Such reification—always a danger in any attempt to explain tacit knowledge—dehumanized individual input by destroying initiative and dampening creativity. This was not Taylor's original intention.

Scientific management in itself is less interesting than the issues it raises. Explaining the context of Thorstein Veblen's surprising view of engineers as the "predestined leaders of a social revolution in America," Layton (1986) contextualizes Taylorism as one of the manifestations of the engineering progressivism Veblen witnessed in the 1910s and 1920s. The engineering progressives held that engineers as a group had assimilated the rationality and objectivity of their discipline and could solve social problems by applying the same methods that were so fruitful in the engineering of materials to the engineering of society. This ideology of engineering helped to define the social role and social responsibilities of engineers as well as to develop their professional consciousness.

Noble (1977) points out that engineers were the first members of industry to advocate the systematic application of scientific methods to management issues. Thus the literature of the management movement between 1880 and 1910 was found exclusively in engineering journals and remained there until well into the 1920s.[4] The fact that Taylorism was wrested from engineers by management suggests that managers eventually realized that control of the workplace should not be left in the hands of an autonomous engineering elite whose interests might not always coincide with their own.[5]

The conflict of interests between engineers and management has continued in arguments over automation and, more recently, in management's attempts to automate engineering drawing with computer-graphics systems.[6] The point I want to emphasize here is that throughout these conflicts, the control of codification has been a major issue. It makes a difference whether codification takes place from the outside—accomplished by programmers, for example—and is then imposed on those using the new tools or whether those who use the programmed machinery are involved in the codification process and the daily adjustment of the tools' use. In the case of drafting, which is discussed in chapter 5, the codification of this crucial level of knowledge into computerized systems resulted in both a loss of prestige for drafters and assumptions that the machine was now the bearer of knowledge because the codification process was attributed to the machine and the company that produced it.

Recently philosophers and sociologists have given increased attention to practice, and this work has heightened awareness that uncodified knowledge may be at least as important as codified knowledge in many types of work. As Bucciarelli (1994) demonstrates, however, engineering education still teaches reductionism. In textbook problems not only are machine parts stripped away to leave problems that can be dealt with mathematically, but the real working world of real engineers—a world filled with negotiations, decision-making under uncertainty, and the mixing of engineering principles with social, political, and financial constraints—also disappears.

Alfred Schutz's (1970) discussion of practical knowledge is helpful here. He notes that our knowledge of the world of daily life is not integrated into a coherent system because our interests shift continually. Even in the goal-oriented arena of organizations, Herbert Simon (1945) found that objective rationality is so difficult, given the multiplicity of variables and their combinations, that managers tend to take the first best choice and discontinue searching for alternatives, which Simon termed "satisficing." Both Bucciarelli (1994) and Vincenti (1990) point out that because of the complexities and uncertainties in the problems of engineering design, participants similarly set limits on the number of options they will entertain, even when they invoke the idealized concept of optimizing, which rationally combines theoretical tools and quantitative data.

Standardization, Codification, and Interests

The historical process of standardization in engineering is permeated with industrial interests, starting with the standardization of materials and then moving to the standardization of people.[7] Issues of licensing, credentialing, and standard setting are settled at state and federal levels and do not usually attract the attention of the lay public. Products, credentials, people, and the educational process are all codified and standardized to foster recognition for an elite body of knowledge and those who have mastered it (Freidson 1986). In part through lobbying for legislation, but more often through their influence on administrative law and the activities of private bodies, the professions are deeply involved in the process of establishing the official standards that structure the production of goods and services. But the important question is, Who creates the standards in the first place? As Freidson points out, in the United States most standards have been *chosen* by government officials rather than *created* by them.[8]

By 1916 an aggressive new Bureau of Standards had firmly established itself as a direct link between government and industry. Materials standardization paved the way for industrial standardization. While shop-culture engineers advocated standardization of screw threads, gear teeth, bolts, nuts, and various machine processes for efficiency, they also believed that the manufacturer who dominated a particular market should determine the standards for that market. The school-culture engineers advocated the creation of a centralized authority, preferably as a part of government, that could set standards to which all industries would have to conform. This was an important issue because the economic power of standard-setting organizations is enormous, capable of ruining businesses whose products do not conform to their specifications.[9]

It is in such standard-setting committees that the professions in general and professional associations in particular can have the most unequivocal influence. Freidson (1986, 204–205) notes that such committees are small and that their work tends to be labeled as technical and therefore remains invisible. In general, the committees have a good deal of freedom to establish their own procedures. Their standards may then be incorporated into legislation or, more commonly, into the rules and regulations or administrative law designed to implement legislation. Standard setting is most often performed by official committees or subcommittees made up of the sorts of people whom Freidson (1986, 205) terms ''the good gray members of the profession.'' Those most influential are often individuals who are deeply involved in the politics of official associations. For material product standards, the committee is the place where those members employed by industry are most likely to be influential in representing their employers' interests. Academics and independent professionals are less often active on committees that set product standards, perhaps because firms are better able to pay the considerable costs of participation while individuals and university employers cannot. In the standardization of people the credentialing system works similarly. The professional associations and the professional schools make the important decisions, and the state merely ratifies their standards.

Noble (1977) notes that standardization was seen as a way to eliminate waste in materials and that it laid a basis for the rationalization of production. When the concept was extended to personnel, it was presented as a way to eliminate wasteful efforts by people through the application of scientific management principles. As industrial expansion took place on a large scale, it was the analytically oriented engineer-managers in the large science-based corporations who pioneered in formulating rationalized procedures and were seen as the most rational and therefore most likely candidates to quantify and order the large conglomerates. Part of this ordering included educating the workforce. While working to standardize scientific and industrial processes, secure corporate command over the patent system, and direct human processes, engineers also set out to create an educational structure that could meet the demands for industrial research personnel. They viewed education as the critical process through which the human parts of the industrial apparatus could be fashioned to specification. "Education for life" in the rhetoric of engineering progressives meant that industrial education would prepare one group for a life of labor, while higher education, especially engineering education, would prepare another group for a life of managing labor. The integration of formal education into the industrial structure weakened the traditional link between managers and managed, separating them by the college campus. Because the large industries that hired most of the engineers were the first to recruit college graduates in significant numbers, corporate engineers also helped fashion systems for industry-university cooperation on matters of curriculum and recruitment.[10]

World War I allowed corporate educational reformers to promote their own industrial objectives in the guise of military expediency. An advisory committee was established to assist colleges and universities in reforming education to help support the war effort. Through its lobbying this committee helped create the War Department Committee on Education and Special Training (CEST), which took charge of American vocational and higher education during the war. The educational work of the military during the war was placed in the hands of corporate education directors and the leading advocates of the corporate reform of engineering education.[11] On October 1, 1918, as the War Department agency for educational matters, CEST actually took charge of all colleges of liberal arts, technology, business, agriculture, medicine, law, pharmacy, dentistry, and veterinary medicine, all graduate schools, and all technical institutes in the United States.[12]

The end of the war ended this experiment in what amounted to corporate control of education. However, corporations and engineers soon made plans to continue their work in another sphere. A civilian version of CEST, the American Council on Education (ACE), emerged from the war as the central agency representing higher education and brought together for the first time the nation's largest education associations. Its leadership consisted of former CEST members, and its top priority was centralization of educational authority comparable to governmental control over education during the war years. The goal was corporate control over education through governmental means (Noble 1977).

Just how far business interests had influenced engineering education was revealed in an investigation of the utilities industry undertaken by the Federal Trade Commission from 1928 to 1933, following a resolution of the U.S. Senate. Hearings revealed the close relationship between the utilities companies and educational institutions. Among other abuses, the FTC discovered that utilities companies were subsidizing utilities courses in colleges, saturating schools with propaganda against public ownership of utilities, and conducting summer schools for faculty members. While the FTC investigation sent a shock wave through educational institutions and the personnel-management agencies of industry, it slowed only momentarily the corporate reform of American education that had been going on for thirty years.

In 1932 a committee of the American Society of Mechanical Engineers (ASME), supported by spokespeople for the other engineering societies and the National Council of State Boards of Engineering Examiners, proposed the formation of a permanent agency to direct all activities relating to the education and professional practice of engineering. While its immediate objective was the formulation and enforcement of "minimum professional standards" to stave off compulsory state licensing of engineers, ultimately the new Engineers Council for Professional Development (ECPD) aimed to control all aspects of professional engineering.[13]

World War II further integrated military interests into engineering, and military influence, like corporate influence, continues to permeate all levels of the profession, from cultural authority to standards and codification. I will not go further into the military influence in engineering other than to note that computer-assisted design also emerged from an alliance among academia, industry, and the military. From the 1950s until about 1970, military funding was important in the development of interactive graphics programs that allow manipulation of pictorial representations. Initial interest in developing computerized graphics capabilities grew out of Massachusetts Institute of Technology research into the numerical control (NC) of machine tools, research that was funded by the U.S. Air Force. The Air Force also paid for research to extend the NC machine tool programming language to capture the geometry of parts at the design stage. Military interest and involvement continues, but since the 1970s commercial markets and interests have taken over the military's monopoly on the development of CAD technology (Arnold 1984).

Conclusion

The codification and standardization of parts, processes, and people are intrinsically tied to the history of the profession of engineering. I have argued that the process and the actors involved in codification are as important, if not more so, than the outcome. But there are other historical dimensions to the codification of engineering knowledge, as well. This chapter has sketched some of the institutional aspects of codification, presenting professionalism as a historic product based on a market project—an attempt to translate one order of resources (specialized knowledge and skills) into another (social and economic rewards) by standardizing knowledge through university curricula; materials, measures, and processes through textbooks; and industry standards through congressional committees. It has also introduced the concepts of primary and secondary codification. The kinds of codification discussed here have been of the latter sort. We now turn to primary codification, which appears to lie in the murky region between tacit and codified knowledge in the realm of engineers' visual culture, the subject of the next chapter.

3

The Visual Culture of Engineers: Drawing, Seeing, and Standardizing Perception

A design instructor at a major engineering school stops abruptly as he checks a student drawing.[1] A form that should read as a cylinder has been represented as a flat surface, encoding that reads like a glaring grammatical error to the experienced designer. The same error shows up on several other drawings by students who learned drafting using CAD (computer assisted drafting) software. These renderings are not simply mistakes: They are the result of differences between paper-world drafting practices and computer-assisted drafting practices. The conventions of drafting—including rules for encoding three-dimensional shapes on a flat plane—developed in paper-world practice. The connection of such practices to the user's socialization and knowledge, along with daily and constant repetition in constructing representations of the world in a given domain—here, the world of engineering drawing—makes practices so powerful that they become interlocked with a particular way of seeing. This is a visual culture (Alpers 1983): a way of seeing that *simultaneously* both *reflects* and *shapes* how members render the world.

The visual culture of engineering is one in which people turn to drawings when asked a design question, like the member of a NASA research and design team who was told "Better go get the drawings" when he tried to describe a part using gestures and an adding-machine tape. It is more than the collaborative visual thinking of two engineers, so deep in discussion of modifications to their surgical instrument design that they sketch together, using one pad of paper and one writing implement, unconsciously passing the pencil back and forth with a coordination suggesting one mind instead of two. The visual culture of engineering is more than the sum of its parts: the practices of sketching and drawing constitute communication in the design world. Other forms of knowledge and communication (verbal, mathematical, experiential, tacit) are built around these representations. Visual representations are so central to engineering design that meetings wait while individuals fetch drawings from their offices or sketch facsimiles on white boards. A shared visual literacy and ability to read encoded meanings can facilitate coordination or can foster conflict in collective projects. Visual objects not only shape the final products of design engineering but also influence the structure of the work and who may participate in it. These situated, collective practices create a visual culture that, in turn, constricts and constructs the literal ability to see or imagine.

This chapter defines and illustrates the visual culture of engineers as situated in practice, delineating the components of engineers' visual culture and its relationship to tacit and experiential knowledge. A historical account traces the development of drafting conventions in the West and the daily practices that have constructed a visual culture not necessarily compatible with the assumptions built into computer graphics-design. By describing the visual literacy that engineers develop in practice and the levels of encoding in engineering drawings, parallels can be drawn between these encoding structures and the structures found in language and artworks. If we can explain how visual representations can function as boundary objects and conscription devices, then perhaps we can suggest why they are such a powerful tool.

Practical Epistemologies and Visual Culture

A visual culture links explicit material experience to a particular way of seeing the world. That engineers think visually is well documented.[2] That their cumulative social environment influences their visual thought has been given less attention. In art history visual culture has been presented as a link between art and daily visual skills that is tied to the material conditions of rendering the world using brush, pen, stylus, pencil, charcoal, crayon, and so on. Citing Alper's (1983) analysis of Dutch painting, Latour (1986, 1990) points out that visual culture is not a metaphorical but a literal material world view—how a culture sees the world and makes it visible. Thus, "A new visual culture redefines both what it is to see and what there is to see" (1986, 9–10). For those engaged with drawing, painting, or mechanical drafting—any means of representing—how they see their world is strongly tied to the learned conventions of rendering it. Engineers who generate and manipulate visual renditions of objects do so using the conventions of drafting. Work habits and experience also play a major part in the construction of engineers' visual culture.

Sketches are the real heart of visual communication. A senior drafter who had been promoted to engineer fought with management for the return of her drawing board, stating, "I can't think without my drawing board." It is where she solves problems. Sketches facilitate both individual thinking and interactive communication. Because they allow these processes to occur simultaneously, they become group thinking tools. Thorough designers continually use sketches, from early drafts talked over with designers and fabricators to rough drawings in margins to clarify an idea (Henderson 1991a).

Sketching is essential to getting ideas across. Designers talk to one another and sketch simultaneously. Some of the best designers talk over and through sketches, not just with one another but also with those who produce the design. In the case studies, we meet the designer who fought for her drafting board to be returned when she was promoted from drafter to engineer and a design engineer who takes his early drafts to the shop floor to sit and talk with the welders and mechanics and clarifies those early drafts by sketching in the margins of official drawings. Bucciarelli (1994, 92) also notes that the act of drawing what he calls *personal sketches* is valuable not so much to produce a design product as to initiate a basis for the author and others to generate a set of more exact plans that will also be modified collectively. This interactive use of sketches and drawings knits together people with shop expertise, such as welders, with designers, tapping individual expertise in piping, electrical circuitry, lubrication systems, and so on, to think through a machine. But cognition must be situated in practice.

Lave (1988) stresses the link between everyday material experience and thought, visual or otherwise. Her work challenges functionalist, conventional views that formal school-based knowledge makes cognitive skills available for generalization and everyday use. Observing the expertise of grocery shoppers, she argues that knowledge in practice is the locus of our most powerful knowledgeability. Lave's practice theory denotes no division between domestic life and work, domestic and public domains, routine maintenance and productive activity, or manual routines and creative mental work. Hutchins (1995) takes a similar approach to navigation, ethnographically documenting its distributed and cultural construction in interaction. Similarly, the visual culture of engineers is not made up of school-learned drafting conventions but rather the everyday practices of sketching, drawing, and drafting that construct their visual culture—a visual culture that in turn constructs what and how design engineers see.

Alpers (1983) documents how artists' attention to surface detail in seventeenth-century Dutch art reveals how their culture saw the world. Italian Renaissance painters gave us grandiose historical scenes rendered as if viewed from a carefully placed window, using perspective to create the illusion of deep space, but seventeenth-century Dutch painters rendered their world viewed as a close-up mirror, reflecting real details of daily life. This included the forms and textures of rich domestic fabrics, reflective vessels, tools and utensils, fruits and flowers, ornaments, and mirrors themselves. This cultural space was constructed by contemporary preoccupations with the camera obscura, which transformed large three-dimensional objects into small two-dimensional surfaces around which onlookers could turn at will. Also influential was the notion of artifice in German astronomer and mathematician Johannes Kepler's account of the retinal image. Unlike the Italian carefully placed window through which one observes grand historical scenes, retinal images are passively observed, like the mirror surface that reflects what is already there. The situatedness of such a visual culture is linked also to other material conditions.

The Dutch were the primary developers of technologies involving mirrors and lenses, including the camera obscura, telescope, and microscope. They were also significantly advanced in cartography. Their descriptive way of seeing is analogous to the way one views a map—an image without a viewer. To see a map is not to look from some imagined window but to see the world in a descriptive format. Latour describes the impact of the interconnectedness of these materialities on ways of seeing for the period and culture as "the new precise scenography that results in a world view which defines at once what is science, what is art and what it is to have a world economy" (Latour 1986,10).

Similarly, Baxandall (1972) points out the relationship between the proliferation of cones and cylinders composing human, architectural, and other forms in fifteenth-century Italian paintings and the obsessive pleasure merchant art patrons took in using such depictions as if they were tangible objects in order to demonstrate their mathematical skills in calculating volumes, area, and ratios. The visual culture here was such that it was as if painters and their patrons saw the world as compositions of perfect geometric components.[3]

Anyone who has undertaken drawing seriously knows that learning to put what the eye beholds onto paper influences the level of detail one literally sees. The art student is trying to recreate an image; she or he sees shape, framing, point of view, shade, texture, and detail in more levels of nuance than the casual observer. This more thorough looking affects mental representations, too, but is easier to trace in actual renditions. Drafters, illustrators, and shop-floor mechanics are used to a way of seeing rooted in practice. If the format is changed, the work becomes foreign and difficult to understand; for example, Westerners tend to think of pictorial renditions within a rectangular frame that recedes according to the rules of perspective. Why? Despite the available variety of shapes, the reproductions we see most often are captured in rectangular shapes, using a lens that creates an illusion of depth. Photographers frame their shots in rectangular lenses, and we in turn see our world recreated in such frames in magazines, on cereal boxes, and in posters. However, Ivins (1953) points out that when the camera was first introduced, people used to viewing engravings saw photographs as a distortion of nature because the world was not neatly arranged in the composition or figures were cut at the edge of the frame. Similarly, the world seen through a fish-eye lens seems distorted to contemporary viewers because within its limitations both the perspective and the frame are circular.

Visual Thought and Spatial and Kinesthetic Intelligence

Engineering documents are intrinsically linked to visual thinking, from earliest conceptual stages to final production. The mind's eye can range from a fully developed realistic representation to a more general sketch or a simple schematic. The image is not static. Its formal aspects (such as size, shape, texture) can be adjusted at will.

Psychologist Howard Gardner (1984) argues that visual image processing is not mere intuition but is indeed a distinct intelligence. According to Gardner the characteristics of spatial intelligence include the

• Ability to recognize instances of the same elements,

• Ability to transform or recognize the transformation of one element into another,

• Capacity to conjure up mental imagery and then transform it, and

• Capacity to produce a graphic likeness of spatial information.

As a sociologist I have some arguments with the more biological–determinist aspects of Gardner's model, but thinking in terms of other ways of knowing is useful. Although Gardner's multi-intelligence model remains controversial even in his own field, other researchers have confirmed the presence of visually oriented thought processes (Pylyshyn 1981, Shepard and Cooper 1982, Kosslyn 1990), though their approach too is overly biological and mentalist. My focus here is how to link the self-evident skill in visual cognition with observable engineering culture and practice.

Such links between thought and practice can be traced in the visual documents so crucial to engineering design. As we move into the world of practice another often overlooked way of knowing becomes apparent—what Gardner terms *kinesthetic intelligence,* otherwise known as *fingertip* or *craft knowledge.*[4] Both spatial and kinesthetic knowledge combine, of course, with practical experience, which uses and enhances them. Together these other ways of knowing are practical epistemologies. Developed in situated practice, they shape the visual culture of engineers. Logicomathematical and verbal skills are not the primary modes of thought responsible for good engineering design, but rather these practical epistemologies of visual and kinesthetic knowledge are situated in the practice of everyday activities.

Engineer and historian Eugene Ferguson (1977, 1985, 1992) points out that the objects of our daily life (carving knives, comfortable chairs, lighting fixtures, and motorcycles) were determined by technologists—craftspeople, designers, inventors, and engineers—using nonscientific, nonverbal modes of thought. Scientificity is overemphasized and even mythologized. Practical visual thinking in design means that the qualities of the objects in the mind of the technologist are not formal verbal or mathematical descriptions. If we are to understand the development of Western technology, we must appreciate visual thought—the determinant of our physical world. Pyramids, cathedrals, and rockets exist not because of geometry, theory of structures, or thermodynamics but because they were first a visual image for those who designed them (Pye 1964). Turnbull (1993) documents how cathedrals like Chartres were built using local knowledge, templates, and string. Vincenti (1990) gives us this telling 1923 quotation from J. D. North, a British engineer and member of the Royal Aeronautical Society:

Aeroplanes are not designed by science, but by art in spite of some pretense and humbug to the contrary. I do not mean to suggest for one moment that engineering can do without science, on the contrary, it stands on the scientific foundations, but there is a big gap between scientific research and the engineering product which has to be bridged by the art of the engineer.[5]

Another way of talking about such nonverbal and nonmathematical ways of knowing is through the concept of tacit knowledge.

Tacit Knowledge

Not all nonverbal knowledge is visual. What Michael Polanyi (1958, 1967) calls *tacit knowledge,* or a personal way of knowing, informs the explicit knowledge characteristic of science. It is present in the creativity of laboratory practice and the passion of discovery. Knowing is action that requires skill: a carpenter, for example, knows how a type of wood must be handled or what type of joint will best serve a given situation. The carpenter knows these things—but may or may not be able to put the knowledge into words. Similarly, Harper (1987) documents a welder's tacit knowledge of the colors of various heated metals as indications of their state of readiness for bonding or tempering. All elements and uses of scientific knowledge are thus vital—not only the formal and the informal but the political and persuasive, the emotive and intangible, and the unspeakable.

The role of tacit knowledge has become a growing concern in practice-oriented studies of science and technology. I am using the term here in the broadest sense—to signify knowledge that is not verbalized, in some cases because it cannot be but in other cases because it may simply be taken for granted or regarded as too trivial to warrant verbalization. The generation or elicitation of all types of tacit knowledge is intrinsically linked to practice. Collins (1974, 46) states that "all types of knowledge, however pure, consist in part of tacit rules which may be impossible to formulate in principle." His study of newcomers' attempts to build lasers using documentary information reveals that even with access to accurate diagrams and blueprints they could not build lasers without having participated in real laser building. Those whose lasers worked made personal visits and telephone calls. Even then some failures occurred when newcomers failed to understand all the relevant parameters.[6]

Hindle (1981) remarks on how fingertip knowledge from Europe informed the growth of so-called American ingenuity. In the sixteenth and seventeenth centuries, German miners, Flemish weavers, and French glass workers and clock makers transferred technologies from the Continent to England. In turn, the English sent Italian silk reelers, Dutch and Polish glass workers, and German sawyers to their new colony in Jamestown, and throughout the colonial period, technology continued to be transferred from an immigrant group of skilled operatives familiar with its mechanisms.

Tacit knowledge, like intuition, is a residual category that encompasses many dimensions of nonverbal knowledge. I focus here on visual or spatial knowledge. Ivins (1953) has pointed out that technological innovation was blocked for ancient Greeks because they lacked a reproductive capacity for pictorial information. Given the Platonic mind-body split, only slaves worked with their hands. However, hands-on knowledge is intrinsic to visual knowledge and thus technological know-how. Because slaves were the power source in ancient times, there was no impetus to invent technology to conserve human energy and similarly no impetus to attempt to capture and convey such knowledge visually.

Practical Epistemologies versus Perspective: Constructing Optical Consistency

Latour (Latour 1987; Latour and Woogar 1979) has given us the concept of *inscriptions*—images that have been extracted from the laboratory to appear later cleaned, redrawn, and displayed as figures in support of a text. They are mobile, presentable, combinable with one another, and immutable. This immutable quality derives from the translation of information with optical consistency with the original source; internal properties of the subject represented are not modified (1986, 7) In visual modes this is accomplished through some device such as the conventions of map making or drafting.

Based on Edgerton (1980), Latour claims that perspective has contributed to optical consistency (1986). But perspective has only limited validity for engineering practice. Although renderings in perspective played a historical role and continue to generate financial and organizational support for design and commercial promotion, these are not design functions. They utilize a different kind of representation and are drawn by illustrators, not designers and engineers—who scornfully refer to them as pretty pictures in the case study examined in the next chapter. The earliest drawings used to transfer technological knowledge across time and space were illustrations and not plans (Baynes and Pugh 1981). Architectural layout drawings of this early period were actually similar to the layout drawings that carry engineering information today. In design, perspectival illustrations play only a supporting role and do not contribute the optical consistency that is crucial to creating an object from a drawing. Indeed, an engineer in the second case study replied to my question regarding the use of perspective by asking, "Would you want a dress pattern in perspective?" His question succinctly points out the immutable quality of plan, profile, and section-view drawings that can be more easily mapped onto the surfaces of raw materials. A perspective drawing would be completely useless in this respect.

Design does not take place through one monolithic rendition. Rather, designers join the immutable information of drawings at centers of calculation (Latour 1986) and collect them into increasing iterations and sets of drawings, in order to hold pieces of information in the absence of the thing itself. The cascade of sketches ultimately made concrete in the final design representations depends on the mobility, stability, and combinability of the representations for extension of the network of participants and information. This contrasts with scientific practice in terms of direction. As Lynch and Woolgar (1988b, 105) point out, in science the use of mere metaphor, similarity, or surface resemblance in representations is regarded as inadequate and is abandoned in favor of "deep," "genetic," or "mathematical" reconstructions of a phenomenon's organization that "penetrate" the depths of the phenomenon by unraveling or dissecting it. Design engineering is going in the other direction: visual representations, here, must capture the process of building up an artifact, requiring visual articulation of a minutely detailed nature to reveal the object's inner structure and workings. But direction is not the only difference. Bucciarelli (1994) points out that although the rational reconstruction of design in terms of the internal logic of the object alone is always possible and reasonable, from the perspective of design as a dynamic social process it is wide of the mark, if not wrong.

How do the practical, situated epistemologies of design engineers work with drafting conventions to facilitate the optical consistency that captures such detail in the transition from large machine to flat representation and back again? Baynes and Pugh (1981) show that until the late eighteenth century no standards of engineering drawing existed. They note that while it is tempting to include certain medieval manuscripts, publications by Renaissance military engineers, and Leonardo da Vinci's notebooks in the prehistory of engineering drawing, these are not design and production drawings but illustrations. The rationalization of sight during the Renaissance enabled quantification of visual information for accurate transfer of visual information. But it was not until the middle of the eighteenth century that France, spurred by war and trade, officially supported applied drawing. The French military engineer Gaspard Monge developed and codified the conventions of descriptive geometry on which the theoretical aspects of modern engineering drawing are based, and a Napoleonic officer brought the concepts to America via the military academy (Booker 1963). Sponsorship of perspective renderings of architecture and technology proliferated, certainly serving to create a center of calculation supported by networks of patrons.

While expensive books with beautiful drawings in perspective proliferated for gentlemanly clientele during this period, what did real production drawings look like? Baynes and Pugh (1981, 32) note that the Royal Navy, "that very commercially minded and middle class organization, since the early years of the century had required that a model and plan for each of its ships should be prepared for the Navy Board." The standardized layout and simple conventions of these ships draughts remained unaltered until the advent of steam in the 1830s. These drawings, as well as other commercial plans from the period, were rendered not in perspective but in profile, section, and plan views, as are production drawings today. A profile view, as the name implies, is straight on at 90 degrees from the side. A plan view is top-down or bird's eye. A section view shows a slice to reveal an interior; like the profile view it is represented from a straight view perpendicular to its center. These are rendered in two dimensions without the distortion of perspective's illusion of three dimensions.

The model thus became available for the total three-dimensional representation. The elegant perspective renderings in expensive books of the period were for gentlemen and not for working designers or builders. Although the first practical application of steam power was made by Thomas Newcomen in 1712, no engineering drawings exist from the period (Baynes and Pugh 1981). Because the depictions of Newcomen engines from the period were drawn by artists and not engineers, several are inaccurate. None of these renditions was in any way connected with production work. Engravers, too, faced problems in depicting machinery, noted in this remark by Newcomen's biographer (Rolt 1963, 107):

It must be emphasized that early pictures of Newcomen engines are not an infallible guide to chronological development. Artists either failed to understand the principles of the valve gear and drew it indistinctly or inaccurately, or else they copied their predecessors' work. Thus the Sutton Nicholls engraving of what purports to be the York Buildings engine shows the buly gear only, which is certainly incorrect.

Edgerton (1980), arguing for the deterministic quality of perspective in the development of Western technology, similarly observes inaccuracies in Chinese copies of Western technical drawings in the sixteenth and seventeenth centuries. He maintains that the decorative conventions of Chinese art did not support the depiction of technology, while in the West perspective and the exploded view had a direct effect on technological development. I disagree.

The missing ingredient in both the Sutton–Nicholls engraving and the erroneous Chinese woodcut prints of Ramelli's windlass pump is tacit knowledge. Actual hands-on understanding of how the machine works is not captured by perspective or any other drawing technique. It is the tacit knowledge of the craftsperson—the practical epistemologies of eye, hand, and situated practice—that gets the job done. Any drawing captures only a small piece of tacit knowledge, even if experts are consulted. Drafting conventions were designed to elucidate more unspoken knowledge through standardization, but those conventions do not arise until the late nineteenth century. Renaissance perspective drawings were also illustrations, rather than production-oriented draughts, meant for gentlemen and perhaps investors but not for engineers or workers.

Newcomen, who made the first viable steam pumping engine in the world, did so without drawings, using techniques that were well established to the point of being almost medieval in their rugged durability (Baynes and Pugh 1981). The after-the-fact drawings of Newcomen's engine, many inaccurate, represent the absorption of his innovation into the scholarship of his day by the use of drawing conventions similar to those of Diderot's *Encyclopédie*. Practical engineers of this period did not use perspective drawing as an everyday part of their work. Similarly, Turnbull (1993) discusses how conventions of templates, string, and geometry were used to construct gothic cathedrals without extensive documents.

When the ability to transfer such kinds of undocumented technological knowledge was lost, explicit drawings became necessary (Hindle 1981). At the end of the nineteenth century, technology became more complicated, and innovation expanded beyond small workshops in which the innovator had direct control. The appearance of engineering drawings as a mature medium in their own right coincided almost too neatly with the establishment of the first factory for the construction of stationary steam engines (Baynes and Pugh 1981). In 1773, Matthew Boulton and James Watt founded their Soho Manufactory in Birmingham to produce modified Newcomen engines. Watt's codification and thus standardization of drawing practice is credited with developing the conventions of drafting. In their scope, style, and direct application to design and manufacture, these codes and standards divided engineering drawing from technical illustration. Maximizing on an unusual background—an apprenticeship in instrument making and a membership in the circle of natural philosophers at Glasgow University—Watt wove together the threads of architectural, technical, scientific, and military draughtsmanship into a practical means for design, development, and production control (Baynes and Pugh 1981).

Drafting Conventions

In contemporary engineering design, Baynes and Pugh note the emergence of the following typology of drawings:

- *Designer's drawings*   Early sketches and notebooks,
- *Project drawings*   Formalized early sketches that show proposals that have been produced according to rules and conventions, often by the drawing office rather than the engineer,
- *Production drawings*   Drawings that display all aspects of design for use in actual production,
- *Presentation and maintenance drawings*   Drawings made after the product is finished for use by the customer, and
- *Technical illustrations*   Illustrations for popular books that use some conventions of engineering drawing.

Historically and in contemporary engineering, perspective may be used in all of these formats except production drawings. However, production drawings are central in the implementation of a design into an actual product. Perspective is used where engineers must interact with others, often in nonengineering networks—in project drawings for funding and organizational support; presentation and maintenance drawings for direct consumers, and technical illustrations for public relations. Although Edgerton (1980) focuses on them, these drawings do not get the design-to-production work done. Those that facilitate fabrication are two-dimensional plan, profile, and section drawings, using drafting conventions and the clear practical epistemology of everyday design work. For engineers, designers, and drafters, perspective drawings have a particular space and function as illustrations. Orthogonal or two-dimensional plan, profile, and section drawings are the acknowledged carriers of engineering production information.

Design engineers and drafters may see their design object in many views in their head but eventually reduce these views to the engineering convention of two dimensions on paper. They do not put down the design from just any angle. The plan or layout is an orthogonal view, seen straight down at 90 degrees, viewed from the top or side. *Orthogonal* means mutually perpendicular. Such views, which are sometimes called simply two-dimensional or 2D, represent the object from a 90 degree or profile view (figure 3.1). By contrast, an isometric view, used conventionally for illustrations, shows two sides at once—one at 30 degrees and the other at 60 degrees (figure 3.2). This is less clear for production purposes because perspective distorts sizes and shapes when showing them at an angle. Isometric views are used in illustrations—as pretty pictures—for maintenance and operation manuals.

These conventions become entrenched in practice. A shop mechanic who was responsible for putting together build books for product construction gave me a good example. Drawings for build books are generated from the actual process of building the first production model. The drawings may be computer-generated or hand-drafted by professional staff or vendor drafters. Problems have arisen with some drawings generated on the graphics system. Carol, one of the mechanical engineers whose responsibility is to assemble the build book, mentioned a particular problem with some illustrations that came through the computer-graphics system:

When I can, I use theirs [CAD-CAM drawings]. But a lot of times the angles and things they use are not sufficient for us. For instance, I've got a seal oil system I was working on for about a year and I did use a lot of CAD drawings, but they were confusing to the people on the floor. See, like I ended up drawing this [points to a drawing similar to figure 4.6, bottom] to tell them the location of where these switches were going to go; whereas originally I had something like this [figure 4.6, top]. And they felt that angle was too, too drastic, and they couldn't read [it] that well. They wanted to look at something and see it. So they threw out my CAD drawing, and I redrew it. And that happens a lot. . . . It's nice to make pretty pictures, but the main, important thing is to make it as simple as you can for these guys 'cause they're the ones we're doing it for: so they can do it right.[7]

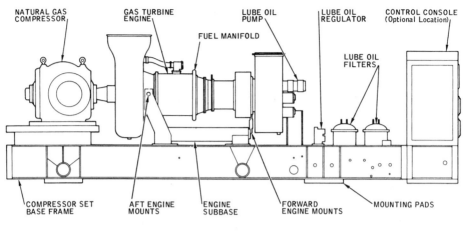

NATURAL GAS COMPRESSOR   GAS TURBINE ENGINE   LUBE OIL PUMP   LUBE OIL REGULATOR   CONTROL CONSOLE (Optional Location)

FUEL MANIFOLD

LUBE OIL FILTERS

COMPRESSOR SET BASE FRAME   AFT ENGINE MOUNTS   ENGINE SUBBASE   FORWARD ENGINE MOUNTS   MOUNTING PADS

SIDE VIEW

CA4

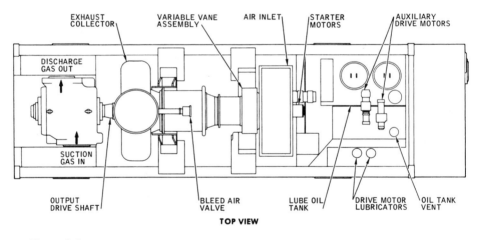

EXHAUST COLLECTOR   VARIABLE VANE ASSEMBLY   AIR INLET   STARTER MOTORS   AUXILIARY DRIVE MOTORS

DISCHARGE GAS OUT

SUCTION GAS IN

OUTPUT DRIVE SHAFT   BLEED AIR VALVE   LUBE OIL TANK   DRIVE MOTOR LUBRICATORS   OIL TANK VENT

TOP VIEW

Figure 3.1
Sample orthographic illustration. From *Selco Technical Publications Illustration Style Guide*. Corporate copyright, pseudonym used to preserve anonymity. Used with permission.

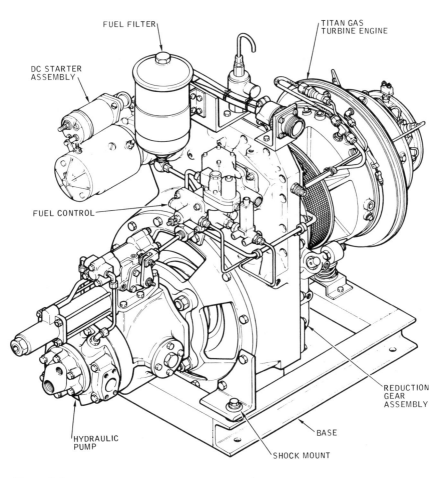

FUEL FILTER

TITAN GAS
TURBINE ENGINE

DC STARTER
ASSEMBLY

FUEL CONTROL

REDUCTION
GEAR
ASSEMBLY

BASE

HYDRAULIC
PUMP

SHOCK MOUNT

Figure 3.2
Sample isometric illustration. From *Selco Technical Publications Illustration Style Guide*. Corporate copyright, pseudonym used to preserve anonymity. Used with permission.

The hand drawing done by the mechanical engineer was simple, without dimensions or drafting skills and no more polished than the crudely drawn facsimile shown in chapter 4. Yet it made its point when the cleaner, computer-generated drawing failed because the crude drawing followed the norms of the visual culture. The mechanical engineer's drawing became part of the build book. Similarly, the packaging design manager reported that designers would not accept computerized representations because they needed orthographic or two-dimensional views and resisted the requirement of the CAD-CAM system that the whole plan had to be developed in perspective before plan and profile drawings could be generated from it. The manager complained:

Designers and draftsmen always have very, very rigorous rules. I'm in real trouble here with the CAD-CAM system because mostly the guys that design are used to producing designs in orthographic representation, and they don't consider a design until that's done, whereas the CAD system really is designed around producing isometric, three-dimensional representations, and then, from that, the CAD system produces plainer drawings that are essentially orthogonal representations. I myself am not a design engineer. . . . You've got a problem here in that the engineer perceives that everybody needs orthogonal drawings to build this thing. I don't think you do—but you try and tell that to some of the engineers, one of them particularly. . . . He gets very, very upset with the whole CAD system because he says, "It's easy enough for you to work in these stupid isometric pictures, but in the end you've gotta have an ortho"—you know, a "proper drawing," as he calls it.

Actually, to design and produce something—whether a turbine engine, a medical instrument, a nuclear power plant or a garment—plans with only one relationship to shape and dimension are needed. This is most clearly and simply accomplished by conventional, dress-pattern-like, two-dimensional representations. Illustrations using perspective create illusion, not exactness by distorting dimensions to give the appearance of recession in space. According to the glass box approach in engineering design, each side of the object is hypothetically traced onto the appropriate side of the box, and then the box is opened flat so that each face is rendered in two planes only. Each side is clear, illustrated in terms that make visual sense with the measurements (figure 3.3). Perspective would present a shorter and longer line both with the same measurement, a confusing state of affairs. Hence at the transfer of information for fabrication, two-dimensional drawings became conventional since they allow the immutable transfer of the information.

While some conventions remain standard in all drafting, others are company specific, such as symbol system sets and lettering. Military drawings have their own standards and conventions. The learning and constant use of such conventions lead to the acquisition of a type of practice-oriented visual literacy—a situated practice. The actual process of drawing is not a school-learned activity that carries across to practice. Engineering designers do not follow textbook protocols. They structure their work in an apparently random fashion, moving from one part of the design to another as the ideas flow (Ullman, Stauffer, and Dietterich

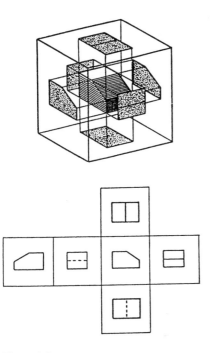

Figure 3.3
Top: The glass box used to define views for orthographic rendering. Bottom: The glass box opened out leads to six basic views. From Booker (1963). Used with permission.

1987). The two engineers mentioned earlier, sketching with one tablet and one pencil, unconsciously sharing the writing implement as part of their conversation concerning a particular design problem, were not exhibiting formal drafting behavior. But they were certainly highly accomplished visual practitioners.

Space will not allow me to cover all the visual conventions of drafting, but I need to mention a few others beyond the use of two-dimensional production drawings. One important formal drafting convention is standardization of the use of line. Variations in the thickness of line are used for specific design designations, with the thickest line assigned to indicate the major components in mechanical engineering (figures 3.4 and 3.5) or to indicate the lines carrying power in electronic engineering (Bucciarelli 1994). Standard curves and shapes are rendered using a template; less standard curves can be traced using a malleable template. Another important convention is the manner by which dimensions are

LINE WEIGHTS	
*PEN SIZE	USE
00	• Object Lines • Graph-Grid Lines • Center Lines • Projection Lines • Dimension Lines • Phantom Lines • Reference Lines
0	• Object Lines • Schematic Component Parts and Boxes • Block Diagrams Boxes • Organization Chart Boxes • Graph "Tick" Marks
1	• Object Lines • Flow Diagrams and Electrical Schematic Flow Lines • Block Diagrams Flow Lines • Organization Chart Flow Lines • Graph Curves
2	• Graph Curves • Graph Borders • Outlining or "Stylizing" to Emphasize Assemblies or Component Parts. See figure 2-9.
3	• Detail Boxes • Outlining or "Stylizing" to Emphasize Assemblies or Component Parts. See figure 2-9.

*Pen sizes No. 00 through No. 1 can be used for object lines.

Pen sizes No. 000 used only when absolutely necessary to define a small part, touch-up, or for rework on reduced art.

Size of original art and final reduction will govern pen size for "stylizing."

Figure 3.4
Chart showing appropriate line weights for different elements in drawings. From *Selco Technical Publications Illustration Style Guide.* Corporate copyright, pseudonym used to preserve anonymity. Used with permission.

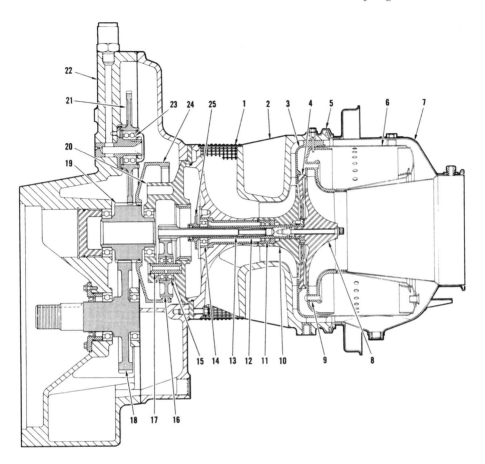

1. Air Inlet Screen Assembly
2. Air Inlet Housing
3. Diffuser
4. Seal Plate Assembly
5. Flange Clamp
6. Combustor Liner Assembly
7. Combustor Housing Assembly
8. Turbine Wheel
9. Turbine Nozzle
10. Compressor Wheel

11. Turbine Roller Bearing
12. Rotor Shaft Assembly
13. Sun Pinion Assembly
14. Turbine Thrust Bearing
15. Ball Bearing
16. Star Gear
17. Star Gear Support Shaft
18. Output Gear
19. Output and Accessory Drive Gear
20. Carrier

21. Accessory Drive Intermediate Gear
22. Reduction Drive Housing
23. Ball Bearing
24. Ring Gear
25. Bearing Retainer and Oil Slinger Nut
●26. Lube Oil Pump Assembly
●27. Oil Pump Drive Gear
●28. Fuel Control Drive Gear
●29. Starter Gear

● NOT SHOWN

Figure 3.5
Sample cross-section illustration showing various line weights. From *Selco Technical Publications Illustration Style Guide*. Corporate copyright, pseudonym used to preserve anonymity. Used with permission.

put onto a drawing, a matter of major importance when design specifications reach the shop. Drafters with shop experience put dimensions onto drawings starting from a point marked zero and consistently proceed with that ordering. This approach allows the shop worker to cut the material without several repositions of the blade. While some conventions remain standard in all drafting, others are company specific, such as particular symbol systems, and lettering norms (figures 3.6 to 3.8). Drawings that must meet military specifications have their own standards and conventions. The learning and constant use of such conventions—formal, informal, or local—leads to the acquisition of a type of visual literacy that is practice oriented. In a *situated practice*, the actual process of drawing is not a school-learned activity that transfers directly from the classroom to practice. Ullman, Stauffer, and Dietterich's 1987 study showed that engineering designers often rely more heavily on the flow of creativity and improvisation than any strictly prescribed method of problem solving. The two engineers in the example given earlier, who sketched together sharing one tablet and pencil, were feeding off each other's ideas, allowing inspiration rather than convention to guide them. The talking sketch (figure 5.6) illustrates how designers employing a less formal method can also exhibit highly accomplished visual literacy in practice.

Visual Literacy

Greenfield (1984) has documented children's acquisition of a visual literacy for television editing techniques. She notes cross-cultural research done on visual literacy among young audiences of *Sesame Street* (Salomon 1979; Salomon and Cohen 1977) that shows that children who grow up watching television have different visual abilities than those who do not. Television watchers recognize that different views or cropped views of a room are indeed the same setting; nontelevision watchers see each view as a different site. Similarly, nontelevision watchers are more likely to identify a long shot and a close-up shot of the same person as two different people. Constant exposure to and interaction with a way of seeing develop skills in visual reading analogous to verbal reading and writing literacy. Because language is tied to culture, the way we speak both reflects and reinforces our cultural outlook and values. Similarly, the visual literacy of engineering designers reflects and informs their visual culture. As with language, not all visual literacies are the same. They may

TYPICAL FLUID POWER GRAPHIC SYMBOLS

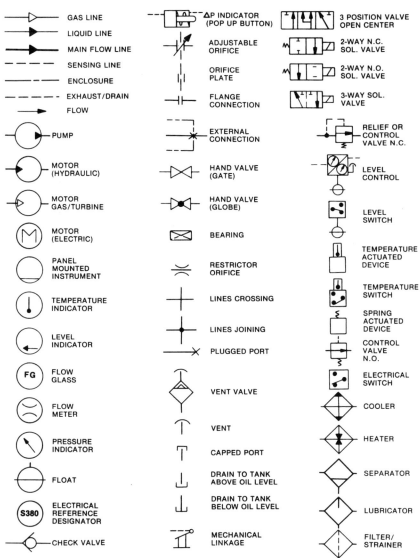

Figure 3.6

Lexicon of graphic symbols for fluid power as used on fluid power diagrams and schematics. From *Selco Technical Publications Illustration Style Guide*. Corporate copyright, pseudonym used to preserve anonymity. Used with permission.

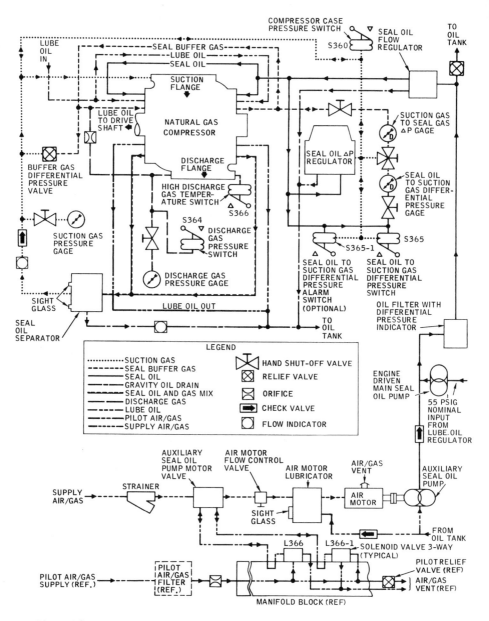

Figure 3.7
Single-line flow diagram illustrating use of lexicon of graphic symbols. From *Selco Technical Publications Illustration Style Guide*. Corporate copyright, pseudonym used to preserve anonymity. Used with permission.

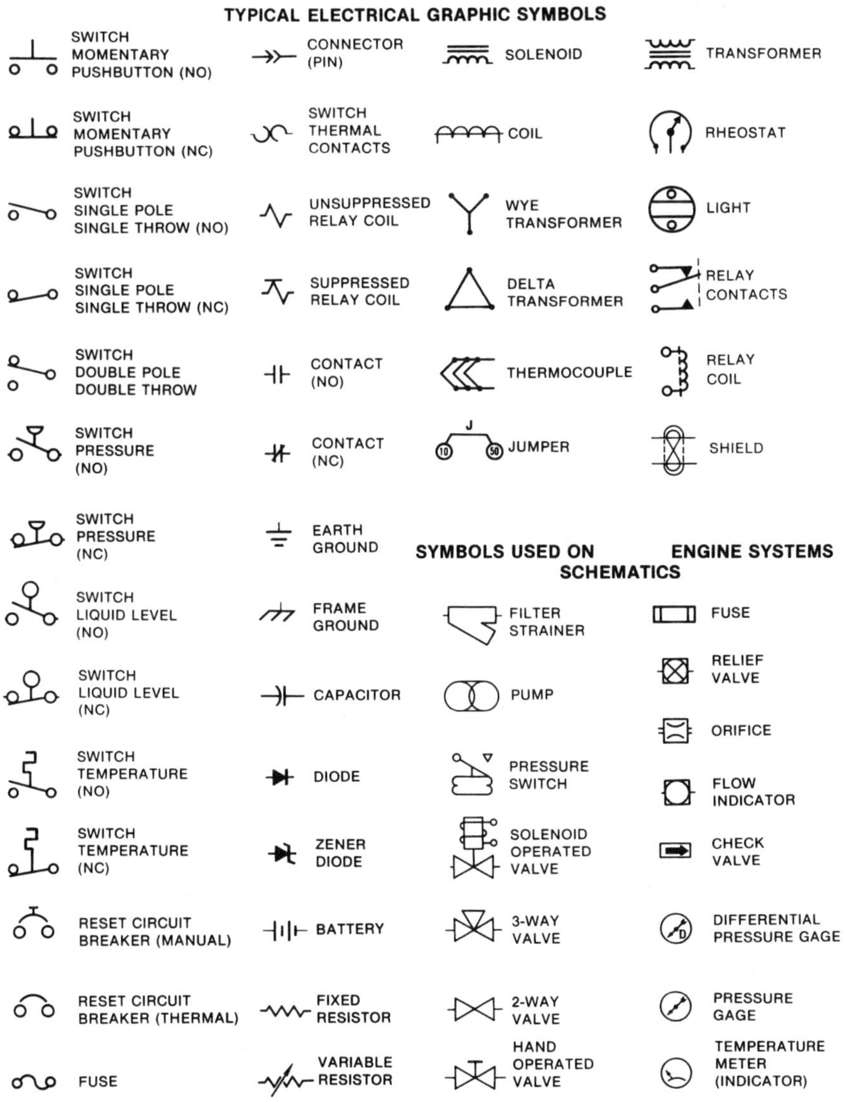

TYPICAL ELECTRICAL GRAPHIC SYMBOLS

SWITCH MOMENTARY PUSHBUTTON (NO)

CONNECTOR (PIN)

SOLENOID

TRANSFORMER

SWITCH MOMENTARY PUSHBUTTON (NC)

SWITCH THERMAL CONTACTS

COIL

RHEOSTAT

SWITCH SINGLE POLE SINGLE THROW (NO)

UNSUPPRESSED RELAY COIL

WYE TRANSFORMER

LIGHT

SWITCH SINGLE POLE SINGLE THROW (NC)

SUPPRESSED RELAY COIL

DELTA TRANSFORMER

RELAY CONTACTS

SWITCH DOUBLE POLE DOUBLE THROW

CONTACT (NO)

THERMOCOUPLE

RELAY COIL

SWITCH PRESSURE (NO)

CONTACT (NC)

JUMPER

SHIELD

SWITCH PRESSURE (NC)

EARTH GROUND

SYMBOLS USED ON SCHEMATICS   ENGINE SYSTEMS

SWITCH LIQUID LEVEL (NO)

FRAME GROUND

FILTER STRAINER

FUSE

SWITCH LIQUID LEVEL (NC)

CAPACITOR

PUMP

RELIEF VALVE

SWITCH TEMPERATURE (NO)

DIODE

PRESSURE SWITCH

ORIFICE

SWITCH TEMPERATURE (NC)

ZENER DIODE

SOLENOID OPERATED VALVE

FLOW INDICATOR

RESET CIRCUIT BREAKER (MANUAL)

BATTERY

3-WAY VALVE

CHECK VALVE

RESET CIRCUIT BREAKER (THERMAL)

FIXED RESISTOR

2-WAY VALVE

DIFFERENTIAL PRESSURE GAGE

FUSE

VARIABLE RESISTOR

HAND OPERATED VALVE

PRESSURE GAGE

TEMPERATURE METER (INDICATOR)

Figure 3.8
Lexicon of graphic symbols used on electrical diagrams, schematics, and engine system schematics. From *Selco Technical Publications Illustration Style Guide*. Corporate copyright pseudonym used to preserve anonymity. Used with permission.

be embedded within one another or read on different levels by different users.

Universal and Restricted Codes

We acquire literacy by learning symbol systems or codes. According to Bernstein (1971), who initiated the concept of *restricted* and *elaborated codes,* restricted codes—such as the syntax and jargon used within social classes or work groups—are generated by social structures and serve to transmit culture. Restricted codes are predictable, simplified and narrow, impersonal, concrete, condensed, neither analytical nor abstract. While restricted codes limit vocabulary and flexibility, they also designate group solidarity. Elaborated codes include unpredictable form, a high level of grammatical organization and verbal selection, relatively explicit meaning, and elaboration of unique experiences, allowing for modifications to suit the listener. Douglas (1970) and Loftus (1988) have applied the concept of elaborated and restricted codes to visual phenomenon. However, Bernstein, Douglas, and Loftus maintain the division between restricted and elaborated codes, all of them seeming to miss the point that codes can be embedded within one another and that speakers and lookers can be multilingual or multisighted.

For Douglas (1970), the New York abstract expressionist artists of the 1950s share a restricted code that is accessible only to those knowledgeable about the New York school art world. She interprets their abstraction as something that could be spelled out in full, the kind of abbreviated form of communication that is implied by Bernstein's concept of the restricted code. She contrasts this restricted code to more recent works in the new realism or superrealism movement, which she sees as an elaborated code with universal accessibility. While she might be correct about the restricted code of the abstract expressionist school, since most art movements start with a small group who share a way of seeing, she is on less firm ground with her interpretation of new realism or superrealism as an elaborated code and her claim that the rebirth of realism is a shift away from the confidence of in-group communication to a more accessible code for the public. She seems not to have considered (1) objects represented in realistic-appearing art can be as symbolically encoded in a restrictive sense as can the intentions of abstract art and that (2) the formal intent of realistic-appearing art—aspects such as composition, texture, brush stroke, tonal dissonances, and harmonies—may be read by an informed audience using the same decoding devices that it uses to decode abstract art but that are unaccessible to the uninformed.

To elaborate on the first point, realistic-appearing art is an illusion produced by marks made in paint, graphite, or other media to resemble something recognizable in the tangible world. Take, for example, Dürer's *Melencolia I* (figure 3.9). This Northern Renaissance engraving appears to represent an adult angel and *puti* (baby angel), both quite glum, surrounded by various tools and objects, geometric shapes, a skeletal dog, and an airborne bat. Though we can recognize most of the objects represented, their relationship to one another is enigmatic without access to the restricted code through which Dürer and his contemporaries interpreted the engraving in 1514. Dürer's writings and Panofsky's scholarship (1945) reveal that melancholy was regarded as a type of human nature. Hence, all the motifs employed in this engraving pertain either to the state of melancholy or the emergence of geometry in painting through the then-new development of perspective in painting. For Panofsky, the disarray of the symbols of the geometric profession (compass, ruler, square, sphere, and trapezoid) jumbled among the tools of the craftsman (plane, saw, hammer, nails) conveys a feeling of discomfort and stagnation. Because melancholy was associated with the mythological personality of the planet Saturn and with artists as proof of their artistic temperament, and because Saturn wielded and bestowed power and was occasionally represented with keys, the deranged and neglected condition of Melancholy's keys and purse indicates a temporary absence of wealth and power. The ignorant infant, almost blindly making meaningless scrawls on his slate, signifies practical skill, which acts but cannot think, while the mature and learned Melancholy signifies theoretical insight, which thinks but cannot act. (Note the Platonic mind-body split, once again.) The dog, obviously suffering neglect, and the bat also belonged to Saturn. Artists and intellectuals of the period, with whom Dürer conversed and corresponded (making an in-group school similar to that of the New York abstract expressionists), would have understood the restricted code of this engraving. We cannot do so without the help of period scholarship, even though we can recognize all its component parts. Thus, the realism of the work is universal on a very limited level.

Embedded Codes

Bernstein's designation of only two possible types of coding is somewhat reified in its dichotomization. Codes, in fact, are embedded in a visual

Figure 3.9
Albrecht Dürer, *Melencolia I,* 1514. Courtesy of the Fogg Art Museum, Harvard
University, Gift of William Gray from the collection of Francis Calley Gray.

representation and can be read on different levels by different viewers. Some of these decodings adhere to the artist's or patron's intent, while others may not. Code systems are recognized by a broad audience in degrees, not as either elaborated or restricted. For example, thirteenth-century artists and patrons could decode the highly restricted code of meanings in an annunciation scene such as the Merode Altarpiece (figure 3.10): the lilies signify the chastity of the Virgin, the roses stand for her charity, the violets for her humility, and the shiny water basin and towel as tribute to Mary as the "vessel most clean" (Janson 1977). At an intermediate level, contemporary art viewers familiar with the Christian tradition probably cannot decode the work on such a specific level but still understand the narrative content of the scene as depicting the angel Gabriel announcing to Mary that she is pregnant with the son of God while remaining a virgin. However, those unfamiliar with Western Christian culture probably see only a long-haired figure with wings facing a preoccupied woman in a richly furnished interior. The ability of the painting to create fully shared cultural meaning depends on more than mere realism: it includes the larger visual culture. The depiction functions on multiple levels: some audience members read all of the coding, and others are able to read less. Realism is not the only criterion, as Douglas (1970) suggests.

This is what I mean by an *embedded code*. For those who understand the code, it is there for deciphering, but the work may be read on other levels as well. I have similarly argued elsewhere that the diverse individuals who support a folk art museum appreciate the same objects at diverse levels of interpretation and understanding. It is the embeddedness of codes—their ability to be read on many levels simultaneously coupled with the various visual literacies of engineering specialists—that allows engineering drawings to serve flexibly as boundary objects and conscription devices.

Boundary Objects and Conscription Devices

Objects that allow members of different groups, stratified or nonstratified, to come together for some common endeavor, though their understandings of the object of their mutual attention may be quite different, have been labeled by Leigh Star (Star 1989, Star and Greisemer 1989) as *boundary objects*. Successful boundary objects exhibit flexibility by allowing for more specific or restricted readings of codes to be embedded within a more universal one. I first recognized this flexibility

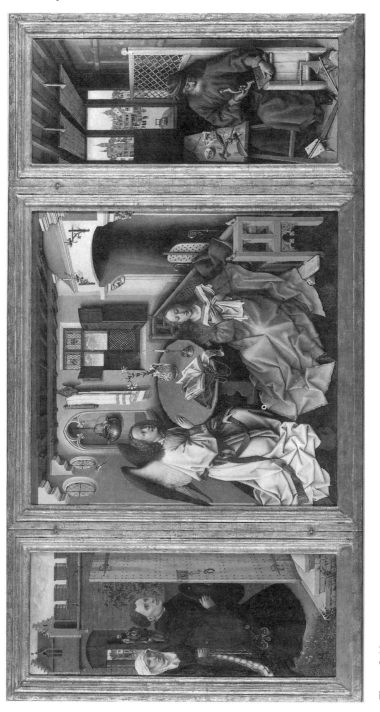

Figure 3.10
The Master of Flaemalle (Roger Campin). *Triptych of the Annunciation.* Courtesy of the Metropolitan Museum of Art, The Cloisters Collection (56.70).

in researching the different appreciation levels for folk art works in a museum's community of supporters. (Star's initial example is a gopher in a natural history museum that has different meanings for groups within that community of supporters.) In both cases one object is understood on multiple levels by diverse groups or individuals. This is particularly important in engineering which both case studies (see chapters 4 and 5) show is a collaborative effort. Vincenti (1990) points out that engineering design is multilevel and hierarchical, and that the design task is divided into many component parts at various levels. Design sketches and drawings are necessarily boundary objects that help integrate various perspectives, such as electrical engineering, fluid flows, structures, and so on. Contested readings can also occur. In the second case study the manufacturing team at a medical optics firm read drawings for a new product as inadequate. Their perceptions were colored by erroneous rumors from field tests and problems encountered on a previous project. In such cases, determining who has the right to participate in constructing and interpreting the boundary object becomes an issue.

Indeed, determining who has access to drawings and who can participate in their construction is also an issue. I have developed the term *conscription device* to accentuate how engineering drawings operate as network-organizing devices and how their creation includes power issues. Conscription devices, a subgroup of inscription devices, enlist group participation and are receptacles of knowledge that is created and adjusted through group interaction with a common goal. To participate at all in the engineering design process, actors must engage one another through visual representations of the conscription device. Participants focus their communications in reference to the visual device. Moreover, early design drawings are guarded from exposure to unwanted input, including input by management, which may otherwise be difficult to ignore.

The focus of conscription devices is process, while the focus of boundary objects is a product. During the design process conscription devices exert a powerful influence. Participants find it difficult to communicate about the design without them. In the absence of visual representations at a meeting, someone will leave to fetch crucial drawings or will sketch a facsimile on a whiteboard or chalkboard (present in all engineering conference rooms) when communication begins to falter.[8]

As a product bearing knowledge that means one thing to some group members who use it and something else to other members, these conscription devices serve simultaneously as inscription devices for capturing information and as boundary objects. Star (1989, 46) describes boundary objects as "objects which are both plastic enough to adapt to local needs and constraints of the several parties employing them, yet robust enough to maintain a common identity across sites." Detail renderings in engineering drawings are one of the tightly focused portions that make up the more flexible whole of a drawing set. For example, the depiction of a welded joint may stand for part of the support structure to the designer and for labor expended to those in the shop. If workers suggest a formation to save welds and if designers incorporate their advice, collective knowledge is captured in the representations as inscriptions. One small part of the welders' tacit knowledge becomes visually represented in the drawing. Hence, the sketch or drawing as a boundary object or inscription conscripts additional work and knowledge from the network interacting with it. While this conscripted knowledge can be tacit or experiential, it can also be other forms of engineering design knowledge such as those described by Vincenti.[9] The sketch or drawing used interactively can serve as a reference and collaboration ground to unite all these various forms of knowledge for negotiation. This concept is more fully explored in the final discussion of the power of visual representations as meta-indexicals in chapter 9.

The practical epistemologies of drafting conventions ensure optical consistency during the process of knowledge accumulation, from large-scale machine to flat rendition and back to machine again. Representational conventions in contemporary engineering incorporate many different techniques to embed information into flat renditions. One technique is an entire lexicon of schematic symbols that designate types and specific functions of component parts of the machine being designed. In this pictographic lexicon, an abstract symbol represents a functional component, and additional visual codes elaborate the basic shape to add specific information. For instance, if two triangles designate a basic valve, the addition of a third triangle can indicate a three-way valve (see figures 3.6, 3.7, and 3.8). The lexicon allows the schematic drawings in which it is employed to remain sufficiently flexible so that engineers can read the coded functions in the layout and understand the interrelations of the various functional components of the whole project. The main concern of an individual engineer may be tightly focused on design aspects in her or his own area of expertise—for example, the electrical components and functions embedded in the drawing for an electrical engineer. Schematic layout drawings utilize this picto-

graphic lexicon to depict the entire relationship of functional components and give detailed information for those working on a specific
aspect of the design. But users have to be able to read the encoding.
Bucciarelli (1994, 95) points out the abstract quality of a section view of a
photoprint machine that has been marked up by an engineer to indicate
potential locations for an additional part. He notes that it is readable
only because those involved share the engineer's same object world and
are very familiar, through experience, with the portions of the machine
it represents and the conventions for rendering them.

I have discussed visual literacy as analogous to reading literacy in terms
of the ability to read symbolic codes. To be literate, one must be able
to decipher a given code as well as manipulate its components in order
to communicate. However, this is constrained by cultural conventions.
The violation of such conventions was the problem raised in the above
example of the students' erroneously coded cylinder. The mistake is interesting because it reveals conflicts in the visual cultures of paper-
trained and electronic-trained designers, exemplifying the link between
practice and a way of seeing.

The visually literate design instructor reads drawings as one might
read a text and is brought up short by an anomalous figure that makes
sense in its context only as a cylinder, but that has been represented by
several students as a flat plane. The glaring error affronts rationality and
breaks drafting conventions. Materially, it is only the omission of a single
line, running across the top opening of the cylinder. But it designates the
difference between a cylinder and a flat plane. For those long initiated in
the visual literacy of drafting, the omission of the line is analogous to a
glaring spelling or grammatical error in a text.

When we compare paper-world and computer-world practices in generating the drawing, it becomes obvious why many design students
trained using computer-graphics systems would leave out the line. The
paper-world practice for drawing the cylinder involves first using a hard
pencil to draw very light guide lines (figure 3.11) to block out the piece
in the correct dimensions. Then the cylinder is sketched within the
guide lines using a darker pencil, and finally the cylinder is inked in
over the pencil drawing. The line across the opening of the cylinder is
already in place from the guide lines and is not likely to be forgotten
since it is essential to the process.

The electronic-world practice does not require guide lines. One types
in coordinates to compose only one side of the cylinder, creating either
a set of rectangles or an L shape (figure 3.12). In electronic drafting, a

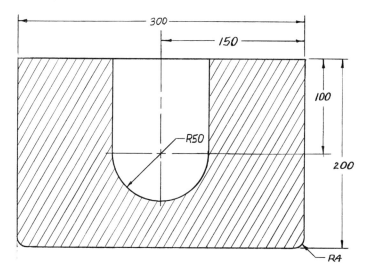

Figure 3.11
Sketch of cylinder as it would be drawn on paper. Note that construction line traces line closing the cylinder opening. Graphics by Robert Bolton.

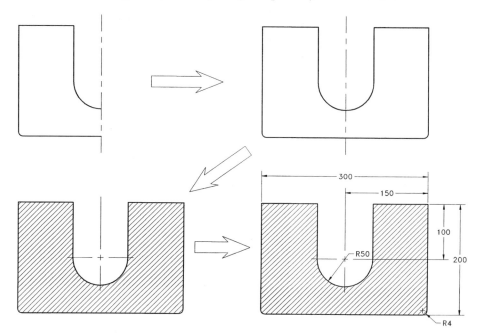

Figure 3.12
Sketch of cylinder as rendered using computer graphics system and mirror function. Note absence of line closing cylinder. Graphics by Robert Bolton.

symmetrical shape can be produced using the mirror function to simply copy one side in reverse, as one might cut a paper folded along its center axis to produce a symmetrical shape. When the half-profile of the cylinder is complete, the drafter tells the computer to use the mirror function to create the other side of the figure, and then the job is done. The figure looks like a cylinder. Nothing suggests it should have a line across the opening. Only a memorized drafting convention, unrelated to the process of creating the figure, might remind the student of the need for the crucial line. The meaning of the line is part of the process, visual culture, and visual literacy of the paper-trained designers. But for computer-trained designers and drafters, it is an arcane memorized detail that is easily forgotten. Young designers trained on graphics software are developing a new visual culture tied to computer-graphics practice, that will influence the way they see and will be different from the visual culture of the paper world.

Visual culture is not static: it can change just as other aspects of culture can change. But when culture changes, it becomes a different kind of knowledge, just as rich, one hopes, as the knowledge it replaces. Zuboff (1988) maintains that paper plant workers develop intellective knowledge as they develop literacy in reading the computerized monitoring of paper pulp and that this replaces former tacit knowledge for gauging the correct pulp viscosity by touch. Modern homemakers use sophisticated washers and microwave ovens for convenience. However, knowledge of special fabrics may demand hand washing, and special-occasion meals may be produced "the way Mom made it." Like embedded codes that may be read on multiple levels by various multivisual readers, computerized design information needs to remain flexible so that users may choose the best use for each application: automation needs to be available for routine, repetitive tasks, but paper-world practice is still necessary for visually thinking, visually analyzing, and visually solving more complex design problems. In the following chapter, these concepts are put into context as a team of design engineers follows local and traditional conventions down what one of them has termed the yellow brick road from design to production of a new turbine engine package.

4

The Yellow Brick Road to Production: Development of a Turbine Engine Package

After a few hours the road began to be rough and the walking grew so difficult that the Scarecrow often stumbled over the yellow bricks, which were here very uneven. Sometimes, indeed, they were broken or missing altogether, leaving holes that Toto jumped across and Dorothy walked around.
—*L. Frank Baum,* The Wizard of Oz *(1900)*

Working and seeing in a visual culture do not necessarily mean that communication among engineers is always perfect.[1] Still needed are the messy interactions that occur in everyday practices and that move new ideas from conception through sketches and drawings to prototypes and production. One of my informants referred to this deceptively linear-appearing process as the "yellow brick road"—suggesting that the design team, like Dorothy and her colleagues, follow a promising path but encounter unexpected happenings along the way. Indeed, this metaphor seems more apt than one that describes an information society that is obsessed with controlling a riverlike information flow. By describing information as if it were a flowing stream from which any unauthorized person along the banks can steal magical properties to use against the rightful owners, this metaphor assumes that the stream must be dammed up and that its borders must be made secure with armed or electronic border patrols.

Information flow is a myth. Information does not flow but rather must be constructed interactively by the human and nonhuman actors involved. Dealing with information is less like floating down a river and more like building a boat. The thesis of this chapter is that the visual communication that takes place between actors in engineering environments is the glue that holds communication together. It allows humans to act collectively toward shared ambitions, to interact with nonhuman actors, and to keep the design and production process close enough to the yellow brick road so that all are not lost by going off in different directions in the woods. Bucciarelli (1994, 111–113) also has noted that the linear flow-charts and block diagrams reproduced in engineering textbooks are intended to establish control but are reductionist and shed little light on how design acts are actually carried out. They contain no hint of the knowledge or resources that participants must bring to the task, the ways in which participants must collaborate, or the symbiotic relationship that exists between the machine and the organization.

Revisiting Conscription Devices, Inscription Devices, and Boundary Objects

The argument presented here uses data collected during participant observation of the construction of a turbine engine package—including actions of the turbine engine that both construct and are constructed by the network of the engineers who must redesign the engine. The empirically informed focus on process challenges the notion of information flow, showing instead the hard work of making tacit knowledge visible so that those involved can engineer a competent design. The cognition distributed throughout the participants is captured in a progression or, to use Latour's (1986) term, a *cascade* of visually oriented inscription devices. As previously mentioned, I have termed these visual representations *conscription devices* in that they enlist the participation of any who would take part in the design or production process. Users must engage these boundary objects to participate at all and do so from different and simultaneous perspectives. The following taxonomy of these visual conscription devices traces the course of group cognitive interaction—from such sources as placemat and napkin sketches, through design renderings and modifications, to documentation drawings used for shop-floor assembly and other versions employed for general production, and finally to illustrated maintenance manuals for customers.

An examination of the roots of the overused term *information flow* is in order here. The word *information* comes from the Latin *informare*—"to give form or shape to." The concept of giving shape to an idea acknowledges that actors must work at shaping and forming ideas through communication among themselves. This is very different from the generation of a disembodied liquid flow that surges on its own but must be controlled by human actors. The word *communication* comes from the Latin *communicare*—"to make common; to share." The problem is not to control a raging tide that might spill over to unauthorized users but rather to control an environment sufficiently so that actors may give shape or form to an idea so that it can be shared. But the act of sharing reconstructs ideas as they are molded from mental images into marks on paper or three-dimensional forms. In an industrial design setting, much of this sharing of ideas takes place in visual modes, such as drawings, sketches, models.

The path from concept to product is therefore not a smooth-flowing stream but is rather, as one senior engineer at this site described it, like the yellow brick road—a designated path of magical promise that is fraught with unforeseen pitfalls, such as malevolent characters and spaces, seductive wrong turns, and the unpredicted actions of the technology itself. Structure—in the form of such entities as hierarchical organization, clearly designated authority, rules, policy procedures, and standardized formats, including computerized databases and computer-graphics systems—is the perceived defense against the demons that lurk along the route in a world that believes it can control them. But I argue that visual representations such as sketches and drawings are the real building blocks of the yellow brick road in this design-to-production endeavor. Furthermore, understanding visual representations as conscription devices and boundary objects allows us to see the glue that holds the entire design process together.

Human and Nonhuman Networks

A further outcome of humans' attempting to control nonhuman actants such as turbine engines is the construction of each by the other, a concept Bruno Latour has proposed in lieu of the fruitless realist versus social construction debate between the science community's view of itself and the social science view of so-called hard science. Latour (1990, 1992) explains a world in which nonhuman actants such as microbes and machines as simple as a traffic bump are constructed and given particular shape and meaning within a specific network. At the same time, these actants participate as actors in the construction of the network environment including the construction of actions that human actors take within it. The following quote shows that the turbine engine also acts. These comments were made by a design manager during a meeting of engineers engaged in designing a new turbine engine package:

And we're still dumping gallons of oil out into the atmosphere that we don't seem to be able to control. . . . Now we have generator sets, Zeus generator sets, sitting out now, which are lubricating northern California. They lubricate the whole area up there. They're dumping fifteen gallons a day out the vent lines. Well, that is a serious problem, and we're spending sixteen thousand dollars a package to overcome it.

The engine referenced here is certainly acting: it is spewing oil smoke into the atmosphere, and the engineers must respond. Spewing smoke is not all this recalcitrant machine does, according to the manager:

I'd like to make sure that I don't have the resonance like I have on the cold-end drive now at thirty hertz. The thing rings like a bell at thirty hertz. I mean it's, it's designed to ring at exactly th' runnin' speed o' the generator. . . . It has got a natural frequency right smack on the nail at thirty hertz. I mean it's bloody awful like that.

The speaker raises the issue that the company's existing turbine engines exhibit at least two problematic actions: they smoke, and they ring like a bell. He also raises the control issue in the first sentence when he says, "We're dumping gallons of oil out into the atmosphere that we don't seem to be able to control." Control is relevant here on several interlocking levels:

1. Control of existing turbine engines that smoke and ring,

2. Control of future engines so they will not smoke or ring, and

3. Control through organized sharing of ideas during the process of creating a new turbine to achieve number 2 from ideas generated in grappling with number 1.

Turbine engines are not designed from scratch. Before a design meeting ever takes place, actors make political and power moves and alliances within the company and in the world of the market. The need for a new engine design is tied to the organization—its struggles, structure, and economic health—as well as the personal politics and ambitions of those involved in its generation. Callon (1986) and Latour (1987) describe this mutual co-construction of the elements necessary for the social support for the new technology and the new technologies support for such a network. Law (1987b) calls the necessary orchestration of these and other discontinuous elements *heterogeneous engineering.* Bijker (1995) terms their successful focus toward a concerted technological outcome a *technological frame.* In all cases the technology is both socially shaped and society shaping.

In the case of Selco, a pseudonym for a midsize, midtech company that builds large turbine engines for industrial applications, some recent historical context is relevant to understanding behaviors within the company during the study. When I conducted the study, the company had been shrinking for almost two decades following, first, the collapse of the aerospace industry and then the collapse of the oil and gas industries. Company managers undertook another reorganization incorporating extensive layoffs in the mid-1980s, a year before my study was conducted. Though Selco had been experiencing some success with co-generational products that produce energy and make by-product heat available for a secondary application, they also had to work to keep competitive with Japanese companies in their market. The design group I worked with was redesigning the package support structure for a new model of one of the company's smaller engines. As Bucciarelli (1994) notes, boundaries and constraints are critical to design. The goal of this redesign project was to increase the turbine package output from 8,000 kilowatts to 10,000 while manufacturing it at a lower cost. The old version sold for $550,000 with 50 percent profitability. The goal for the new version was a 20 percent reduction in the $225,000 cost. I refer to this project as the Zeus Mark IV. The name, changed here for anonymity, refers to the size and capacity of the engine. Selco also builds three larger engines, all with Hellenistic names retained from the aerospace era. The Mark number refers to consecutive designs of a given engine, with the Mark III being the previous design designation.

The Mutual Construction of Engine and Environment

Although information flow may be a myth, controlling the flow of people who might have access to information is a pervasive practice in industry. This was very noticeable throughout all my fieldwork and particularly so at this first site. Selco, like most industrial production companies, is a high-security establishment. Entrance into the buildings is monitored by receptionists in the office buildings and by uniformed guards in the industrial areas. That receptionists and guards do the same job is illustrated by the practice of having uniformed security replace the receptionists at breaks. Employees leaving the building open their briefcases voluntarily for these gatekeepers to show they are taking nothing unwarranted from the premises. Permanent employees must wear photo identification cards prominently displayed at all times while on company property. Temporary employees wear badges that are color-coded to identify expiration dates set at six-month intervals. Both groups carry a magnetized card that allows them to open the back entrance to the office building and the gate in the chain link fence between the office parking lot and the industrial yard, a path often traversed by engineers and others going from their offices to the shop floor or cafeteria, which is located on the industrial side. The card and guard system controls the movement of people and thus access to information by visibly marking insiders and outsiders. Visitors must sign a log stating their name, company affiliation, purpose of visit, and Selco contact person as well as time of arrival before being escorted into the building. On leaving they must return their temporary paper visitor's card and sign out with the time of their departure. As a visitor I was accompanied at all times by the employee I was interviewing, but when I joined the technical writers group, I was assigned a color-coded temporary ID card and a magnetized door card giving me access to the offices and plant.

My position as a technical writer grew out of my interest in information-access control. When I interviewed the head of Selco's computer-graphics section, whom I had reached through University of California alumni referrals, I asked whether I could observe a project unfold from design conception to machine production and thus track ideas conveyed in a visual format throughout the process. The computer-graphics manager referred me to his boss, the manager of packaging design, who complained about the engine that smoked and rang like a bell. The packaging design manager said he was interested in my project because he saw communication as problematic among engineers. In a position to control access to information himself, especially to outsiders, he proposed a trade: I would become a member of the design team working on the Zeus Mark IV, and in exchange I would become the technical writer for the project. This verbal agreement was more a mark of the manager's horse-trader style of doing business than a real contract or industry norm. As he put it, if I wrote anything for them, the company came out ahead; if I did not, nothing was lost, since I would obtain my research data, and what I subsequently wrote about might be of use to the company. The agreement allowed him to justify to his own superiors both my presence and the time I took from workers. It also served another purpose. As a member of the technical writers group, I was placed into the hierarchy of the organization and thus fell under its stratified systems of formal rules and informal practices. I soon found the boundaries of access to information in my new position.

I was curious about seeing the company's training video tapes, which familiarized new employees with the technical workings of turbine engines. These tapes were made in-house by a team of video technicians and technical instructors and were accompanied by a workbook with fill-in-the-blank questions and machine illustrations. The workbooks illustrated various components of the turbine engine and its internal flows of heat and fluids, and the video incorporated numerous animated sequences showing the workings of the turbine engine in coded colors along with full color views of actual engines. The two epitomized the importance of visual communication for understanding the rudiments of technical production, even for employees not specifically involved in engineering or design.

I obtained a list of the tapes with their official file numbers from a fellow technical writer, telephoned the appropriate office, and identified myself as a new employee in the technical writers group. The person at the other end of the line, however, would not discuss the tapes but insisted that I ask my supervisor not only which tapes I could view but when, where, and if I could see them at all. My call the next day after a fellow worker helped me match code numbers and titles to make my request more specific generated no cooperation in the person at the other end of the line. Finally, I identified myself as a sociologist doing research at the company and was greeted with fawning helpfulness: the tapes were available in an office just down the hall. This incident illustrates both Merton's ritualist who forgets goals through microscopic attention to means and the low status and inefficiency of the new employee who does not yet know how to work the system. This instance of control arose not from company policy or security but rather from a misguided tinkering with the organization to make it run more "efficiently." As a consequence of the company's extensive layoffs the year prior, employees became secretive about their work because they feared that another employee or technology could replace them. The tech writers complained that their hardest job was collecting information they needed to write the start-up and operation and maintenance manuals that went to customers. According to fellow writers and the technical publications supervisor, there was no official reason why the woman on the phone would not give me access to the tapes, but she was known to be a difficult person to work with, in an atmosphere in which everyone was afraid of losing their job due to company layoffs.

I found similar parallels in other interactions in my role as a tech writer. The other writers were friendly and helpful with sharing verbal information and written technical materials. However, when I suggested that I have access to the word processing files for the technical manuals that were to be updated, even my closest coworker adamantly refused, insisting I would be of more help gathering information from the engineers than actually writing. Thus, she maintained control over the tangible work output and the credit received for it. Interestingly, as the resident sociologist and only woman on the design team, I was a welcome visitor in the engineers' cubicle offices. They willingly took time from their work to answer my questions, complain about problems, and tell me stories about turbine engines, the company, or themselves. However, in my role as tech writer, when I sent them requests for updates on new information for the operation and maintenance manuals, the paperwork sat on their desks for weeks on end. The paperwork that would eventually make all this known to the customers who bought the product was of lower priority than the exciting design process—both in their minds and in the information disbursal system set up by company policy. Nor could I gain access through standard authorized request forms to drawings that the engineers had released weeks previously; the drawings had not yet been processed through the company bureaucracy. Complaints about access to the drawings were often raised by support personnel from manufacturing and other departments at design meetings so that engineers often loaned their own drawings or personally made copies of them for others rather than wait for copies to make their way through the bureaucratic system. Access to information therefore could be influenced as easily by personal priority, whim, or status as by bureaucratic red tape, workload, and priority policies.

Modular Space, Modular Work, Modular Engine Packages

In other cases, boundaries were more consistent. Engineers differentiate between the actual turbine engine and the system that supports it. The designers I worked with were in the Packaging Division, which creates the fluids, fuel, and electrical systems along with the structural frame that contains and supports these systems and the engine. One underlying design concept for the new engine was to make the supporting systems modular or self-contained so that parts could be easily removed for maintenance and repair work without dismantling other sections of the machine. This was to be done using cheaper and lighter materials to keep costs down and profits up, while maintaining the company's reputation for the production of very high-quality, heavy-duty turbine engines. This modular engine design could be seen as a simultaneous construction of the modulization of the company itself in that different parts of the design job were done by different departments within the company. The engine design group did their part and then passed the job over to packaging. One packaging engineer worked on fluid systems, another on electrical design, and another on the skid or structural support that would hold the assemblage of parts. This is not surprising, of course, since a large part of the control revolution has been the attempt to achieve interchangeability of parts and people. Here, constant tinkering was necessary on both the organizational and the technical levels to work toward such a goal.

A case in point is revealed by comparing two of the company sites within the same city. Selco maintains three sites in South Bay City. One is used for parts storage. The other two facilities, both of which are production sites, provide the telling contrast. The older facility, called *downtown*, by workers, is near the bay in a section of the navy city that has long been dominated by old-line aerospace and military contractors, though the growing tourist industry is "Disney-fying" areas nearest to the water. Selco's pre-World War II buildings now accommodate an industry as sagging and battered as the wooden frame buildings, corrugated steel fencing, and painted-out warehouse windows of the plant. Uniformed security personnel guard access beyond the eight-foot chain link gates, and engineers work in odd-shaped spaces reminiscent of old military barracks. Their office cubicles are tucked in nooks and crannies around the production buildings. Workspaces forced to conform to new human needs show signs of years of tinkering—functionless doors, awkward halls and entrances, mismatched ceiling tiles, and scars from wall surgery.

In contrast, the leased office building and plant in suburban Sun-baked Mesa, about twenty minutes away, is a model of banal planning. The view from the conference room window reveals carbon-copy, manicured landscaping and mirrored office buildings redundantly reflecting Bauhaus international geometry. Inside, packaging design and technical publications departments are housed in a large maze of tiny cubicles that are segmented by five-foot high partitions and covered with dust-brown carpeting. In the technical publications section, four people share each cubicle, and each engineer is assigned a cubicle that allows only enough room to turn around between desk and drafting board. The short walls and open ceiling space give an illusion of spaciousness, but the halls are so narrow that two people can barely pass one another without brushing. Even though two persons talking in the hall make it difficult for a third to pass, many conversations take place at a wide spot where two hallways intersect. Pinned to the drab fuzzy partition at this intersection point is a small card proclaiming the spot "the crossroads." The title is more than a practical joke: it is a cynical commentary on truncated communication and its solution. The technical publications assistant manager told me that moving into the new cubicled space made him feel as if he had lost touch with all his people. Informal communication was further hampered by new austerity measures after the layoffs: lunch was cut to half an hour, and coffee breaks could no longer be taken in the cafeteria but instead had to be taken in the cubicles since a small kitchen adjacent to the cubicled workroom provided a constant coffee supply. The assistant manager's solution was to linger in the hallways or around the crossroads from time to time to keep in touch with other workers and with what is happening in the company and to facilitate the sharing and shaping of ideas that used to take place across desks in the workspace or during coffee breaks.

The obvious parallels here are between fragmenting the workspace and fragmenting the work and the workers' responsibilities. For design engineers the new site meant removal from working with others in one large room to working in separate cubicles that isolate and fragment the human components of the company, modularizing them in the same way the components of the new engine are designed as replaceable units. This attempt at interchangeability of human parts, not far removed from classic Taylorism, can also be seen in the structural adjustments in the company along with the structure of the workload for engineers in packaging design. A senior engineer described the overall structure of the company as having changed from horizontal to pyramidal (he called it "Christmas tree") and incorporating more layers of management, hired in what another engineer called the "M.B.A. '70s," referring to the plethora of business management graduates hired in that era. The senior engineer pointed out that these new middle managers lacked practical experience and were more tied to paperwork and an ideal of what should be rather than hands-on knowledge of what is. He added that many of these newer managers had worked their way up from sales and not from the design or production side of the company. He said that in the past the product engineer had coordinated every aspect of an engine package from design to "out the door." Now, he said, the person in that position just handled the paperwork, and any follow-through between designs and production occurred because the designer does it:

At one time everybody was in a pool, and various product engineers would draw from this pool to do whatever project they were assigned to do. As the evolution of the product became bigger and bigger and came into more of a standard product, that started to go away, and we were divided into various groups. At one time, it was the packaging group, but then the packaging group was broken into more specialties like generators and compressors. And then eventually, about three or four years ago, they totally divided everything up—have a packaging group for generators, a packaging group for compressors, and a predesign group that will take care of any new business coming in. When they did this, it was done in a way: "You're gonna sell generators, it's up to you to sell generators any way you can." . . .

So for the people in engineering, it meant a different structure. . . . It meant being assigned to a product line and more specifically packaging a product line and sometimes even doing different sizes of generator. You were in a position where you were responsible for three product lines of generators. You'd have a sales order for a Neptune one time, and a Zeus another time. So you were required to become more diversified in what you had to do.

As Cap, the senior engineer, points out, this change has been generated by the change in the turbine actors. The expanded product line—more and different sizes of turbines, more varied packaging, and increased volume—together necessitated a change in the project-by-project team assignments. Up until the late 1970s each engine and its package was built to suit each customer. The same team designed and built additions like the generator set, special air filters for deserts, or oil heaters for arctic climates. In the past ten years most of the innovation has been in the packaging of the turbine rather than in the turbine engine itself. In the mid-1970s Selco began packaging its turbines like a car with expensive options. As Cap put it, you can get the basic vanilla unit and order predefined condiments from a wish book of available options offered by the company based on past experience of modification requests from field users. The $16,000 pollution arrester complained of in the opening quotation is an example. The change to a standard product and increased production meant a change in the work structure for the package design engineers. They were no longer drawn from a large pool to work on one whole project from the ground up, start to finish. Their time and assignments became fragmented. Each design engineer became designated as a specialist assigned to one area of packaging, working only on certain applications such as "generator sets"—but as many as six at once, all of different size and design.

Each design team member is assigned a piece but not the total package. Concentration of effort can change literally by the hour. Cap describes the now-abandoned project structure:

When you get onto a project, you know what you're going to be doing from day to day, and it's a . . . you do a design, and you follow through over six weeks or six months or something. It's, ah, it's pretty much of a straight line. You, you start and draw your charts and one thing and another and you've got goals and the whole thing. You know what I mean. So it's really pretty structured.

He contrasts this to the current fragmented structure of work under the product designation:

But under product I, I'm—When you're working product, ah, you can have six jobs going at the same time. Any one of those can, ah, change at any given day. . . . For instance, Kelly West can call up and say, "We got troubles with this," or "The customer just called in," or "One said the water injection unit on this product won't do." And that changes everything else new 'cause they're number three on your list, and you're working on number six at the time. So you drop number six and go running up to number three and take care of whatever has to be done there. And then the next customer calls in and say, ah, "We need our installation drawing in three days 'cause whoever's going to do the installation is back there and they need to know the footprints" and so on. . . . So when you're workin' product, that is, it is not nearly as structured a venture of what you're doing from day to day.

Cap goes on to describe the product structure of work as less personal as well as more hectic for the design engineer:

All of a sudden you're puttin' out fires. You got four fires goin' all at the same time. You're putting out the biggest one. It's, ah, when you're working product, it can get a lot more hectic. It can change literally by the hour, and that, you just have to be more versatile: what you're doing there is take care of one of those fires.

Working on six projects at once may run smoothly for some projects as they proceed down what Cap calls the yellow brick road toward production. But as Cap implies, to assume that things run smoothly down the yellow brick road all the time is wishing for wizardry. As an example, he describes a worst-case day that happens all too often:

For instance, on a given day it could be, ah, an order will come and we'll get a requirement from a customer and . . . find that they have three old frames out in the yard there. Somebody's canceled on them. So the first thing, we go call up and find out they want to cash in three weeks, a real special order. This has happened before. [I] say, "Can we use the old stuff that's out there?" and they say, "I don't know." [I ask] "What kind of generator? What do we have to do?" and they'll come back pretty soon and say, ah, "Hey, here's the generator we're going to use, and it is this," and I'll say, "Okay." And I'll draw a bunch of stuff and say, "Gee, you know we can use it but I have to modify, modify it by doin' these sort of things. And so pick one and talk to structures."

So you just pick a piece of paper, go out there, and make some sketches, and give it to them, and say, "We'll give you the drawings later." And then about the time you come back in, there'll be somebody'll bug you after the bell and say, ah, "We just got a call for some CAD drawing in Wauwatosa from somebody saying they need their newest layout drawing three days early because some reason or they're going to be in town Monday." So their people are gonna be in town, so we have to get it into CADS. We'll send it. Okay, call up structures, and we'll be able to give them their sketches and time to do that. Okay, that's fine. So you go over and start and turn out the installation drawing, run over, take it t' CADS while they're—. It's night and day. They're not the same thing at all. So jump over and get that done. So sometime, in the meantime, somebody else comes in with another order on a—. When you have six projects going at the same time, which can happen, and depending on which one comes in has the precedent at the time. Ah, sometimes we get three of them with all the same precedents, [laugh] so [laugh] you gotta pick and choose. So that's where it gets more and more hectic, you know.

Such fragmentation of time and attention is not uncommon in industry. Bucciarelli describes how a designer endures similar distraction while she is trying to develop a computer model of a photovoltaic desalination plant. She has to put up with her office being moved, missing data, emergencies on other projects, mismatched formats, and other "fires to put out" (Bucciarelli 1994, 55–61).

Wizards, Witches, Fires, and Red Faces

Note that the structure of Cap's speech is like the structure of the fragmented work. He interrupts himself so that his sentence is left unfinished as he starts another one, just as the work he must leave untended is left unfinished to start another project when he says, "Take it t' CADS while they're—. It's night and day" and again, "somebody else comes in with another order on a—. When you have six projects. . . ." He continues:

Or sometimes there'll be, you know, all sorts of, oh, things that come up unexpectedly. For instance, they'll have a package out there, a generator set, at the moment, setting out on the floor, and they'll call you up and say, ah, "This thing won't go together, and the generator won't fit on there. We can't align the generator. We've got problems mechanically out there." You go out there, you'll look at it and say, "gosh, it should go together." So we go over all the numbers then you start backtracking to find out why it won't go together. One time we found out that the, ah, the generator—. The gear box manufacturer had messed up, put the holes in the wrong place, and the alignment pins are wrong. And so he was in England, so we call up England. He said, "We're absolutely right." And [he] said, "We've never had a mistake before." And (laugh) so you're callin' the guy a—. So [I say], "Can you send us a verification?" and big long silence. (laugh) [He] said, "No." (laugh) So anyway, so what we had to do there, long package with a gear box comes off and then we take it and we put it on the machine. . . . That's off three-tenths of an inch or something. The manufacturer said, "Well, we messed up. This is three-tenths of an inch off." "You want the gear box back, or shall we go ahead and fix it?" And he's in England, and we need the gear box on the line in three days, so (laugh) . . . so we have to make arrangements. [I] go over and make arrangements to get the thing machined properly. That can turn into about a three-day arm-waver when they pack 'em in and it's supposed to go together, supposed to go together just like—like stacking boxes. And that doesn't happen sometimes, and that gets interesting. . . .

We change vendors on something, you go to put it together, and now the shop is all mad at you, said, "This here won't go together. You have a lousy coupling." You go out and start to look at all the bits and pieces there. It's all supposed to go together. The changes aren't on the drawing to change or anything, so you start looking at the parts. And so pretty soon you find out that they didn't do something on that part that the other vendor had done. The vendor will come in and say, . . . "We did it exactly like you planned." And I'll take them out and show them they didn't do it exactly like (laugh) like (laugh) I planned. Another red face, and they go running all over. Or sometimes they do it exactly like our print, and our print was wrong, and there are red faces over here. (laugh) So . . . that's more product line, gettin' products out. . . .

If any of these combinations fall in a three- or four-day period it (laugh) gets—. We're trying to put out a lot of, a lot of fires, and most of the time where it really gets more hectic is when these sort of things happen and we have a cash date on. A cash date, what happens: we have a base that comes out, and then everything comes in baskets, if you will, and shows up on the floor. We have a big pile of parts, is what it comes to. That's the cash date for all the parts arrive on the floor. And, ah, between the cash date and the day it hits the test cell and it leaves here is a pretty rigid, rigid schedule. A lot of time it's a very rigid schedule. And a lot of time there'll be penalties that'll amount to ten, fifteen thousand dollars a day if we don't make it. And if we hit one of these projects, and about the third day down the line, the third day into the assembly, ah, we get one of the problems, and it has to be in the test cell in, you know, a week and a half, then it really gets to be a lot of push and shove. There's a lot of things that have to be done in a hurry and, ah, if it can't, it's a vendor problem or it's our fault. . . . That's the first determined. . . . And then the fix that has to be put in place. They have to be done in a very timely manner. . . . It can really get hectic over the short period of time, and there's a lot involved that when it hits the end of the line. Well, you know, it's ten thousand dollars a day. There's a lot of people that—(laugh) are come lookin' down their nose: "What are you doin' about it today?" So, again, when you look at the product, a number of products, it's more demanding or can be more demanding.

However, despite the "arm waving" and "red faces" of high-demand periods, it is obvious there is satisfaction in this work when things are busy, especially in a company that experienced extensive layoffs. In contrast, Cap adds that when design and production run according to plan, he is trapped in his cubicle, "down in the doldrums" of boring paperwork:

[It] can go on for months at a time where it's really down in the doldrums, too— where you're not really doing that much. All you're doing is making sure that everything you said should be done is being done. You go over, yeah, it's being done. And you go back and—(laugh). . . . Everything is going along real well, and it's more paperwork than it is design work. You go through a long period of that. That's a lot less interesting. Pretty soon you start hoping there's a problem, (laugh) just to pick up the interest, as long as it isn't a big problem.

As can be seen by Cap's descriptions, the production even of a standard version "vanilla" turbine package can be fraught with problems of various sorts on the trip down the yellow brick road that leads the turbine engine eventually out the door to the Emerald City of customers and profits. This quest-fraught-with-trials characterization of industrial design and production is a vivid contrast to the concept of information flow, which suggests a smooth liquidlike process without bumps, flaws, or interruptions much less charlatan wizards, crafty witches, threatening forests, or seductive poppy fields in forms such as incompetent vendors who drill holes in wrong places, impatient customers who suddenly show up and want their drawings, foul-ups in production in your own shop or structural changes that promise miracles but make daily worklife into either hectic "arm waving" or mind numbing "doldrums." What holds together the yellow brick road so that design engineers arrive at the Emerald City with their turbine engine intact? The design engineer's vision of the product, his or her colleagues' contributions to the turbine engine package, and the coordinated effort that produce it are bound together by the object that serves as a container for visual knowledge and a vehicle for collaboration. Just as the assistant manager technical writer circumvents the fragmentation of workspace and interaction by using the hallways as a conduit for informal exchange of knowledge, visual representations, in formats ranging from napkin sketches to formal layouts and schematics, glue together the engineers' world of ideas and action despite structural fragmentation of the workload. Thus, each of these visual conscription devices serves as a building block in the yellow brick road from conception through production.

Modes of Visual Knowledge: A Taxonomy of Conscription Devices

As noted earlier, the concept of conscription devices builds on Latour's use of *inscriptions,* images that are extracted from the laboratory to appear later cleaned, redrawn, and displayed as figures in support of a text and that display the characteristics of being mobile, presentable, readable, combinable with one another and immutable (Latour 1986, 1987). The immutable quality of an inscription refers to its translation into a format of optical consistency with the original source so that the internal properties of the subject represented are not modified (1986, 7). This is accomplished through some device such as the conventions of map-making or drafting.

Engineering designs are built up through a cascade of representations and rerepresentations to construct a final design, as Latour (1986) suggests is the case with other cascades of inscriptions. However, then the socially constructed design in which things were turned into paper must be used to turn paper back into things. Since in most cases someone other than the designers must physically as well as socially construct the object that the designers have collaboratively planned, the conventions of drafting discussed in the previous chapter are employed to make explicit the knowledge required to produce the object. They fail to make all nonverbal knowledge explicit, but this does not alter the intention, as with the visits to the shop floor employed by Cap, as discussed above. Such actions acknowledge the importance of tacit knowledge, which too often can be dismissed as the trivial craftsmanship of the laboratory in science or intentionally withheld from other scientists in order to maintain an edge in areas of new research. Conscientious engineering designers who consult with shop workers as well as with other designers succeed in incorporating at least some levels of tacit knowledge in their designs.

As previously noted, conscription devices, a subgroup of inscription devices, enlist group participation and are receptacles of knowledge created and adjusted through group interaction aimed toward a common goal. Participants simultaneously engage one another and the design issues, interact through the visual representation of the conscription device, and focus their attention and their communications with one another in reference to it. The conscriptive quality of these visual representations is so strong that participants find it difficult to communicate about the design without them. If the drawings are not brought to a meeting of those involved with the design, someone will sketch a facsimile when communication begins to falter or a team member will actually leave the meeting to fetch the crucial drawings so group members will be able to understand one another. Also discussed earlier was the role of conscription device as boundary objects that allow multiple levels of reading by different audiences. For example, at this site, the depiction of a welded joint may stand for part of the support structure to the designer and stand for labor expended to those in the shop. If the designer consults with workers who suggest a formation that will save welds and then incorporates the advice, collective knowledge is captured in the design. One small part of the welders' tacit knowledge becomes visually represented in the drawing. Hence, the flexibility of the sketch or drawing as a boundary object facilitates in enlisting the aid and the knowledge of additional participants.

At the same time, the enlisting drawing must remain sufficiently flexible for such transfer of knowledge to take place. Latour cites the historic contributions of linear perspective and descriptive geometry to engineering drafting in maintaining optical consistency and thus facilitating the change of scale from large machine to flat renditions (1986, 28). Though he is mistaken about the role of perspective in production drawings, following Edgerton (discussed in the previous chapter), he is correct regarding descriptive geometry and the conventions of drafting. Indeed, representational conventions in contemporary engineering incorporate even more techniques to embed precise information into flat renditions such as an entire lexicon of schematic symbols that designate types and specific functions of component parts of the machine being designed. This example is a pictographic lexicon in which an abstract symbol represents a functional component and additional visual codes elaborate the basic shape to give more specific information. For instance, if two triangles designate a basic valve, the addition of a "T" shape can indicate a valve with bleed port (figures 4.1 and 4.2). The lexicon allows the schematic drawings in which it is employed to remain sufficiently flexible so that engineers can read the coded functions in the layout and understand the interrelations of the various functional components of the whole project, though the main concern of an individual engineer may be more tightly focused on the aspects of the design applicable to her or his own area of expertise—for example, the electrical components and functions for an electrical engineer.

The ability of schematic layout drawings that utilize this pictographic lexicon to picture the entire relationship of functional components while giving detailed information for those working on a specific aspect of the design nicely illustrates Star's definition of boundary objects as

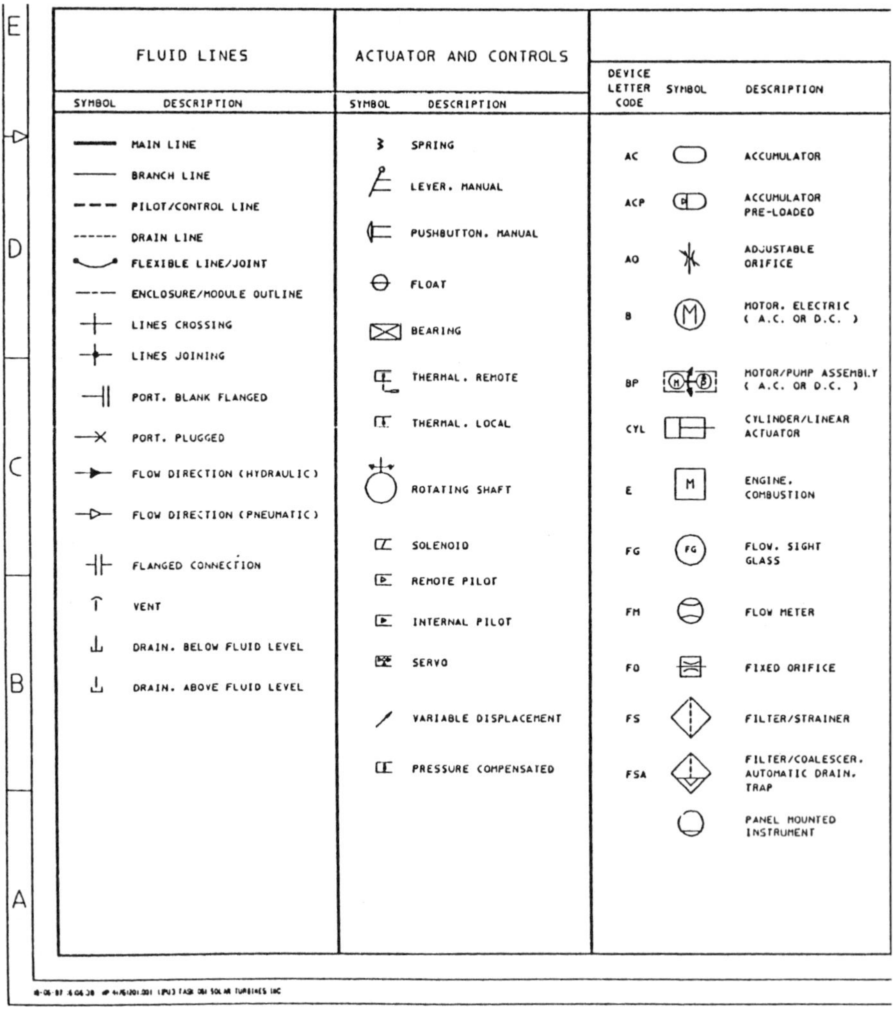

FLUID LINES		ACTUATOR AND CONTROLS		DEVICE LETTER CODE	SYMBOL	DESCRIPTION
SYMBOL	DESCRIPTION	SYMBOL	DESCRIPTION			
▬▬▬	MAIN LINE	⌇	SPRING	AC	⬭	ACCUMULATOR
———	BRANCH LINE	⌁	LEVER, MANUAL	ACP	⬭	ACCUMULATOR PRE-LOADED
– – –	PILOT/CONTROL LINE	⟜	PUSHBUTTON, MANUAL	AO	✳	ADJUSTABLE ORIFICE
------	DRAIN LINE	⊖	FLOAT	B	Ⓜ	MOTOR, ELECTRIC (A.C. OR D.C.)
⌒	FLEXIBLE LINE/JOINT	⊠	BEARING	BP		MOTOR/PUMP ASSEMBLY (A.C. OR D.C.)
— —	ENCLOSURE/MODULE OUTLINE					
┼	LINES CROSSING		THERMAL, REMOTE	CYL	⬚	CYLINDER/LINEAR ACTUATOR
╈	LINES JOINING		THERMAL, LOCAL			
⊣‖	PORT, BLANK FLANGED			E	Ⓜ	ENGINE, COMBUSTION
⟶✕	PORT, PLUGGED	⊙	ROTATING SHAFT			
⟶▶	FLOW DIRECTION (HYDRAULIC)		SOLENOID	FG	Ⓕⓖ	FLOW, SIGHT GLASS
⟶▷	FLOW DIRECTION (PNEUMATIC)		REMOTE PILOT			
╫	FLANGED CONNECTION		INTERNAL PILOT	FM	⊘	FLOW METER
⊤	VENT		SERVO	FO	⊠	FIXED ORIFICE
⊥	DRAIN, BELOW FLUID LEVEL	╱	VARIABLE DISPLACEMENT	FS	◇	FILTER/STRAINER
⊥	DRAIN, ABOVE FLUID LEVEL		PRESSURE COMPENSATED	FSA	◈	FILTER/COALESCER, AUTOMATIC DRAIN, TRAP
					⬭	PANEL MOUNTED INSTRUMENT

Figure 4.1
Lexicon of lube oil schematic for a turbine engine. Corporate copyright. Used with permission.

FLUID DEVICES

DEVICE LETTER CODE	SYMBOL	DESCRIPTION	DEVICE LETTER CODE	SYMBOL	DESCRIPTION	DEVICE LETTER CODE	SYMBOL	DESCRIPTION
H		HEATER	P		PUMP, FLUID	TC		THERMOCOUPLE
HX		HEAT EXCHANGER	P		PUMP, FLUID DRIVEN (INTEGRAL UNIT) (PROPORTIONAL)	TCV		TEMPERATURE CONTROL VALVE (PROPORTIONAL)
L		SOLENOID VALVE, 2 POSITION (TYP.)	PCV		VALVE, PRESSURE CONTROL (N.O.) (PROPORTIONAL)	TI		TEMPERATURE INDICATOR
L		SOLENOID VALVE, 3 POSITION (TYP.)	PCV		VALVE, PRESSURE CONTROL (N.C.)	TW		THERMOWELL
L		SERVO VALVE (PROPORTIONAL)	PDI		DIFFERENTIAL PRESSURE INDICATOR	VCH		VALVE, CHECK
LG		LEVEL SIGHT GLASS	PDI		DIFFERENTIAL PRESSURE INDICATOR (POP-UP BUTTON)	VCS		VALVE, CHECK SPRING LOADED
LI		LEVEL INDICATOR	PI		PRESSURE INDICATOR	VF		VALVE, FLOAT CONTROLLED
LU		LUBRICATOR	PT		FLOW SENSOR	VGL		VALVE GAS LOADER (PROPORTIONAL)
M		MOTOR, FLUID	RT		RTD	VH		VALVE, HAND
M		MOTOR, PNEUMATIC	S		SWITCH, PRESSURE	VI		VALVE, INSTRUMENT WITH BLEED PORT
MU		MUFFLER/SILENCER	S		SWITCH, TEMPERATURE	VLF		VALVE, LIQUID FUEL REGULATING (PROPORTIONAL)
N		NOZZLE, FLUID	S		SWITCH, LEVEL	VM		VALVE, FLOW MODULATING (PROPORTIONAL)
			S		SWITCH, DIFFERENTIAL PRESSURE	VR		VALVE, RELIEF
			TB		THERMAL BARRIER	VT		VALVE, TRANSFER
						V2P		VALVE, 2 POSITION
						V3P		VALVE, 3 POSITION
						Z		PRECIPITATOR, ELECTROSTATIC

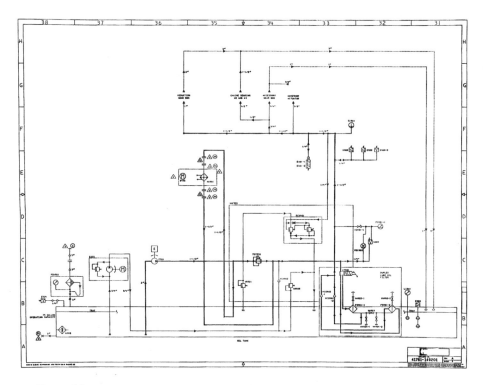

Figure 4.2
Sample lube oil schematic for a turbine engine. Opposite page shows an enlargement of the bottom right corner. Corporate copyright. Used with permission.

"weakly structured in common use" and "strongly structured in individual–site use" (1989, 46). At the same time, this very definition reveals the problem of CAD-CAM at Selco, where the set of official fixed drawings is so strongly structured, because of its many interlocking computer databases, that its ability to function as a boundary object is impaired. I discuss this in more detail shortly. First, it is necessary to examine the building blocks of distributive cognition more closely, so I return to the story of the Zeus Mark IV.

Models
The Mark IV project officially started with a model. This model was not built in the Selco model shop but rather by Rod, a middle-management "M.B.A. type," to use company jargon, who subsequently became the project engineer for the project because of particular interest. This indi-

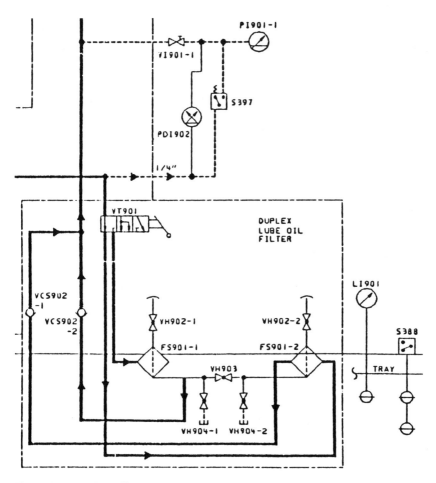

Figure 4.2 (continued)

vidual built models as a hobby—not kit models but scale-model miniatures from plastic parts that resemble real industrial raw materials such as I-beams and sheet metal. The project engineer to-be had actually applied to apprentice in the company model shop in the past, but the company encouraged no apprentices there, and the official model shop continued to be run by a gentleman near retirement. A diversion to discuss the status of the model maker and the model shop is in order here.

Despite the industrial excitement and marketing promotions for computer modeling, at the time of this study in the late 1980s Selco still employed a lone model builder whose shop was equipped with miniature

I-beams and other small plastic parts that had to be configured into shape just as the full-size steel parts had to be made ready for shop assembly. Though the model builder was near retirement, he complained that the company would not allow him to train a replacement, although he was constantly approached by eager volunteers wanting to learn his job. He also complained that when he goes to modeling shows now he finds fewer and fewer model builders and useful workshops and more computer modeling displays. When I first interviewed him in 1984, he was planning to retire in six months and to be hired back on a contract basis to build models. At my last visit in the late 1980s, he was still working for Selco. Doubtless the company foresaw that its need for building tangible models would decline as three-dimensional computer modeling became more prevalent throughout industry. These years of transition when both types of models were in use provide an opportunity for understanding the characteristics of tangible plastic models for later comparison with three-dimensional computer modeling, which I do in the broader comparisons of chapter 5.

Models were used at Selco to make a first approximation of a design and to communicate its appearance and virtues to management in a tangible format. A model previously built for this purpose also was used on the shop floor to give workers a three-dimensional representation of the overall appearance of a new package. This is not trivial. Informants told me that parts sometimes are installed upside down or otherwise incorrectly when a new project is being assembled for the first time. A model may also be used to delineate complicated piping configurations that supply lubricants and fuels to various parts of the engine package. The exact layout of the piping is not included in the designer's original drawings, so the first time a package is built, decisions for piping layout are made on the shop floor. Models can be used to give shop workers a general impression of how the assembled package should look. Models are also made of single-production, custom packages, since they will differ from the norm. Models built for shop-floor illustration are also often used as sales tools to give the customer a concrete idea of the product. Thus, models qualify as boundary objects that represent different meanings and uses for various groups—representation of a future product idea, configuration of piping installation formations, a customized product, or a sales tool, depending on the individual viewer's interests.

The model's three-dimensional, scaled form ensures a high level of Latour's requirement of immutability for inscription devices in that viewers see a plastic form in space rather than a two-dimensional drawing that they must convert into three-dimensional form in their mind's eye. At the same time, their plastic format limits them from combining with other information as sketches do so well. They are different from prototypes that occur interspersed between series of sketches and can therefore be modified and adjusted. But that discussion comes in the next chapter. Models produced at Selco were not working models and hence did not show the internal construction of the machine. Designers at Selco point out that since the models are only first approximations, they can be misleading in details—such as the flow pattern of piping that appears correct from an external view but could run into interference in its path inside the machine. The solid modeling offered by computer–graphics salespeople would seem to offer the immutability of plastic models with the detail of interior drawings, but no one in the packaging division was proficient in three-dimensional computer modeling for new designs. Computer-graphics programs were used by drafters and detailers and not by the senior design engineers who did conceptual design at Selco. Drafters entered hand-drawn copies into the CAD-CAM system after design engineers completed their designs. But I am getting ahead of my story.

The model for the Zeus Mark IV was generated out of conversations between Cap, the senior engineer known as the guru of piping, and Rod, who built the model and later became project engineer. The new design combined some the best aspects of previous designs along with new ideas for modulization. Before the model could be built, these ideas had to be conveyed through rough sketches accompanied by conversation.

Sketches: The Heart of Visual Communication

As noted in chapter 2, sketches are the real heart of visual communication and are the most important carriers of visual knowledge. They serve both as an individual thinking tool and as an interactive communication tool. The flexibility of sketches to serve in both these capacities allows these two functions to overlap when designers collaborate so that sketches become group thinking tools that facilitate distributed cognition. Engineering drawings and sketches do not capture the full array of tacit knowledge used in practice, but like indexicals, they stand for or point to more complex stocks of tacit knowledge and provide a loose, broad frame for them so that individuals and groups may index different information or knowledge from the same representation or portion of it.[2] I am including here not only quick sketches used to convey information between people or used by designers to jot down ideas in visual notes but also early drafts of formal drawings that are used by designers as an interactive tool in that they may be altered or corrected by designers other than the person who drew them. Sketching is connected with thoughts that are in visual format. As an interactive tool, sketches are the most direct way that an engineer can contribute to a colleague's conceptualization of an idea—by giving form to concepts pictured in her or his own mind.

The importance of sketches as individual thinking tools is illustrated by Sharon, a design engineer who said that when she was promoted from drafter to engineer her drafting board was taken away.[3] She asked for it back, stating, "I can't think without my drafting board." Sharon describes the way drawing helps her think:

Just all of a sudden there's a time when you just start laying it down. It's like, ah, . . . when you do drafting by hand. A lot of the thought process comes when you're drawing it out. So it's like you can't think. . . . You can't sit there and wait until you've got it up here because it comes through drawing it. And as soon as you start drawing it, you have ideas and changes. You're erasing it and improving it. And they say the best designers start drawing right from the beginning. That's where they make all their mistakes. . . . They're drawing it out there. They're looking at it. They're visually checking it and improving upon it. . . .

You're just getting a feeling for it. You know, you're just trying to get a size, trying to understand it—to get an understanding of the system, almost building up your self-confidence so you know the task you've been assigned. And when you have the feeling, you want to know how does it work, too, because you've been given all this stuff, and you don't just hook it up. And when you've kind of satisfied yourself that you understand it sufficiently, that you have your data, that you know where you can go to get more data, you start in. And, ah, because then, as soon as you have something, you can take it to someone else and say, "Look, this is what I have. How can I improve on it?" or "Did you have a problem in this area too?" Without the drawing, you know, it's just talk. But when you start laying it out, that's where, you know, 80 percent of the problems come out—when you start drawing it and you start realizing what you've got.

Sharon is delineating several purposes that sketches serve. As an individual thinking tool, the rendition of the visual image onto paper is analogous to the accomplishments of the written word. A physical picture, like writing, captures an idea, allowing it to traverse time and space independent of its author. The tangible form of the idea also allows for critical analysis. As Sharon states, "As soon as you start drawing it, you have ideas and changes. You're erasing it and improving it," and later, "But when you start laying it out, that's where, you know, 80 percent of the problems come out." In the early stages this is a rapid process during which the sketch serves like notes or the outline for a piece of writing. Sharon notes, "They say the best designers start drawing right from the beginning." Like note taking, quick sketching preserves fleeting ideas before they disappear from memory. The sketch captures them in a concrete format and form—literally giving ideas a potential shape that is a first approximation of the actual form objects will take. Sketching also serves another purpose, which is to better understand the parameters of the project itself: "You're just trying to get a size, trying to understand it . . . to get an understanding of the system, almost building up your self-confidence so you know the task you've been assigned."

Sharon illustrates the sketch's use both as an individual thinking tool and as an interactive communication tool showing its flexibility from one to the other as she states, "As soon as you have something, you can take it to someone else and say, 'Look, this is what I have. How can I improve on it?' or 'Did you have a problem in this area too?'" She adds:

You've always got something in hand. You never get—I don't think you ever get two designers who just sit down and just talk. It's "give me a pencil, and I'll explain." Everybody draws sketches to each other, and you know, "This is what I'm trying to do here. Let me show you."

At the interactive level, sketches facilitate communication to further refine ideas. As boundary objects serving to facilitate shared cognition concerning the packaging of a turbine engine, sketches capture pertinent knowledge from many sources. Cap, the senior designer on the Mark IV, points out that visual communication facilitates communication not only between design engineers but also between designers and those in the production cycle:

When you're doing the mechanical design work, you have to be very aware of all the people that are interfacing and need to work with your design problems. You've got to leave room for the electrical people to put their stuff. You have to know how to put a package together so it, ah, can be built in the cheapest, most inexpensive way that you can possibly do it. An' to do that you are in contact a good portion of the time with other departments, especially, ah, eh, structures, which actually weld the thing together. Ah, and usually you go over and ask them, you know, "Here's what I'd like to do. Can you see any special problems with it? . . ."

I usually take my layouts right after I've started, after I do my initial layouts, before they're ever dimensioned or, you know, just be a frame layout. Then I'll go over, and I'll sit and talk with them and say, "Look, eh, here's what I wanna do, and here's the types of material I wanna use. Here's what I've gotta have cut out. . . ." I know most of the things they do, and I know most of the things they can and can't do. And it's just a matter of, "Is this the best possible way to do this particular thing?" And usually it's peripheral. But a lot of times they'll say, ah, "Well, gee, if you did this, over here, this would save one weld."

If the designer consults with workers who suggest a formation that will save welds and then incorporates the advice, collective knowledge is captured in the design. One small part of the welder's tacit knowledge comes to be represented visually in the drawing. Hence, the flexibility of the sketch or drawing as a boundary object helps enlist the aid and knowledge of additional participants. On a cognitive level, Hutchins (1991) suggests that the cognitive properties of groups may differ from those of individuals and thus may depend on the social organization of individual cognitive capabilities.

Such collaboration also makes for a more efficient, cost-productive design as well as probably contributing to better relations between the design team and the producers—not a trivial matter, as we see in the second case study. Cap's dependence on sketches after the layout is finished and the project moves to prototype production is a further example of both the flexibility and clarity of sketches in filling in tacit knowledge in their role as boundary objects. The conventions discussed in the previous chapter have been developed to make explicit the knowledge required to produce a machine. That they fail to make all nonverbal knowledge explicit does not diminish the importance of tacit knowledge, even though scientists may dismiss tacit knowledge as the trivial craftsmanship of the laboratory, mystify it as art or magic, or intentionally withhold it from other scientists to maintain an edge in areas of new research.[4]

As to format, detailed visual information for production drawings is initially provided only in a layout format—a top-down view that shows the placement of parts like a map and in profile and section views (figures 4.3 and 4.4). Depictions of how the final project should look overall and the specifics of small detail areas are not available on the shop floor during the assembly of a first package. (When the project goes into standard production, a *build book* or *traveler*, generated out of the process of the first assembly and integrated with assembly instructions from parts vendors, is available for the shop. I discuss these later.) Because Cap knows from experience that the initial representation can easily be misinterpreted, Cap is on the shop floor during the first assembly of his designs and provides additional visual information when needed. Referring to the two-dimensional layouts from which the initial assembly workers must build the first prototype, Cap acknowledges, "They still may not make as much sense in one area, so you circle a line on it and make a little hand sketch off to one side there to clarify the thing." Pointing to a vague area in a layout he says,

For instance, like where it's black here in this area, they see a dotted line coming down and say, "What's really going on there?" I'm liable to just go off with the pencil and go an' just make a little hand sketch off to one side—just to make it clearer to the guys doing it at the time—but it wouldn't be as a permanent anything.

While not all designers follow through as thoroughly as Cap does, his use of visual communication through hand sketches facilitates both the early design process during its conceptual stages and the actual realization of the design during its final production. He moves back from the weak structure of the layout drawing to the strength of its building blocks—sketches—to fill in the site-specific detail.

The informal visual communication of sketching is essential to getting ideas across. Sharon says that you never "get two designers who just sit down and just talk," adding, "Everybody draws sketches to each other." Cap points out that a meeting with people in structures, where the package is actually built, "works wrong if you can't communicate." He takes his drawings to "sit and talk" so others understand what he is trying to do. Likewise, he draws sketches for workers so they can understand what he has in his mind that may not be clear in their minds from looking only at layout drawings. This interactive use of sketches as conscription devices and boundary objects knits together the input of people with expertise from shop experience (such as the welders in the Structures Division) with the combined knowledge of various designers, tapping individual expertise in piping, electrical circuitry, lubrication systems, and so on to socially construct a machine built from distributed cognition.

Official Production Drawings

The most complex conscription device is the fixed drawing—the official document that is credited with conveying visual information for a given project. Although called a drawing, the fixed drawing is actually a set or several sets of drawings that become official carriers of information when they are released by the project engineer. *Releasing* a drawing means it is in its final form and can be distributed to the various user groups. However, it is widely used as a communication tool before the point that

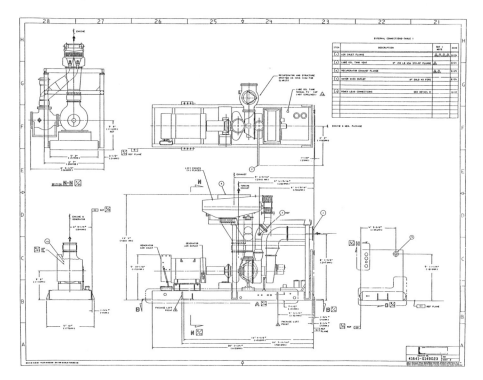

Figure 4.3
Sample plan, profile, and section for a turbine engine. Opposite page shows an enlargement of the middle section. Corporate copyright. Used with permission.

it becomes an official document, as has been discussed under the subject of sketches. The set that makes up the drawing consists not only of the layout drawings of the object itself. It also includes an installation layout, detail views, electrical diagrams, schematics (which are maps of the functional components of the machine), and also parts lists (figures 4.2 to 4.5). The drawings are read like maps. Each part has a letter and number code so it can be easily found within the drawing. This indexing is done in the same manner that streets or towns are found on a road map by locating the intersection of the number and letter grid lines.

The formal drawing serves not only in the designing and fabrication stages of the product but in marketing, sales, inventory control, and accounting. Every minute part of the final product is indexed on the drawings and matched to a part number kept in a computer file. This multifunctioning of the fixed drawing means it serves as a boundary object of much larger scope than the sketches from which it is built. It

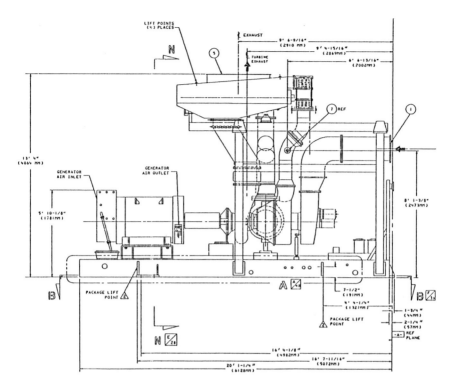

Figure 4.3 (continued)

encompasses many more boundaries in that it is integrated with other computerized databases: the parts indexed to the drawing are also part of inventory and sales records. This is too much overlap for one boundary object because it hampers the flexibility asset of boundary objects as weakly structured in common use. At Selco, the interlocking databases of cost accounting and inventory control overloaded the boundary object and made it appear inflexible so that employees became intimidated by it and looked for ways to work around it. I discuss this in more detail in the section on the use of CAD-CAM at Selco.

First, it is necessary to look at another set of conscription devices. I have illustrated how sketches and fixed drawings conscript participants in their role as boundary objects to facilitate the shared cognition of the small group involved in the initial design process. For the building process to become routine, another level of detail in visual information is needed when the product watched over by an attentive designer moves from design prototype to standard production. This set of conscription

28   27   26   25

REFER. DESIG.	DWG. ZONE	COMPONENT PART NO.	COMPONENT DESCRIPTION	FUNCTION	COMPONENT SETTING	SYSTEM NORMAL OR OPERATING DESIGN VALUE	NOTES
B321	C37	143925-1	PUMP, LUBE OIL	PRE/POST LUBE OIL PUMP, MOTOR AND RELIEF VALVE ASSEMBLY	RELIEF VALVE CRACKING PRESSURE 11-15 PSIG	FLOW: 11.6 GPM (44 L/MIN) WITH 24 VDC AND 1750 RPM	
FS901-1	B32	917646C91	FILTER, DUPLEX	FILTER MAIN LUBE OIL SUPPLY	5 MICRON		
FS901-2	B32	917646C91	FILTER, DUPLEX	FILTER MAIN LUBE OIL SUPPLY	5 MICRON		
FSA901	C38	948269C3 (TRAP 948279C1)	SEPARATOR DIFFERENTIAL PRESSURE GAGE AND TRAP	SEPARATION OF LUBE OIL FROM AIR	DIFFERENTIAL PRESSURE GAGE 0-10" H2O (0-2.5 KpaD)		
H390	A38	120168-1	HEATER, LUBE OIL TANK	LUBE OIL SYSTEM PREHEAT		120VAC, 3KW, 1Ø	
LI901	B32	945720C1	INDICATOR, LEVEL	INDICATION OF LIQUID LEVEL IN LUBE OIL TANK	"FULL" READING = 75 GALLONS 6.5 - 7.5 INCHES FROM TOP OF LUBE OIL TANK. "1/2" READING = 38 GALLONS 10.25 - 11.25 INCHES FROM TOP OF LUBE OIL TANK.		
P901	C36	919381C1	PUMP, MAIN LUBE OIL	SINGLE ELEMENT, ENGINE DRIVEN MAIN LUBE OIL PUMP		MAIN LUBE OIL PUMP PUMP "A" (P901) 50 GPM (189 L/MIN) AT 100 PSIG (600 kPa) AND 1915 RPM	
PCV901	D34	913739C2	VALVE, PRESSURE CONTROL	CONTROL OF LUBE OIL PRESSURE-MAIN	RELIEVES AT 50-55 PSIG PILOT PRESSURE	FLOW NOMINAL IS 55 GPM MAXIMUM IS 150 GPM	
PDI902	C32		GAGE, DIFFERENTIAL PRESSURE	INDICATION OF FILTER ΔP		0-75PSIG (0-500 kPa)	
PI901-1	H37	111220-4	GAGE, PRESSURE	LUBE OIL HEADER PRESSURE READOUT		0-160 PSIG	
S322	G37	120621-9	SWITCH, PRESSURE	PRE-LUBE PRESSURE PERMISSIVE	6.0 ± 0.5 PSIG INCREASING 4.0 ±0.025 PSIG DECREASING	DPDT	
S324	B38		SWITCH, PRESSURE	HIGH TANK VENT PRESSURE ALARM	8.5±0.4" H2O (2115±100 kPa) INCREASING 7.5±0.6" H2O (1866±150 kPa) DECREASING	SPDT	
S380	G37	120623-5	SWITCH, PRESSURE	LOW LUBE OIL PRESSURE SHUTDOWN	35 ± 1.2 PSIG INCREASING 25 ±0.6 PSIG DECREASING	SPDT	
S381-1 S381-2	E33	946847C2	SWITCH, TEMPERATURE	HIGH LUBE OIL (HEADER) TEMPERATURE	SHUTDOWN 178-180°F ALARM 168-170°F	DPDT	
S388	B32	915695C1	SWITCH, LEVEL	LUBE OIL TANK LOW LEVEL ALARM LUBE OIL TANK LOW LEVEL SHUTDOWN	ALARM AT 55 GALLONS SHUTDOWN AT 30 GALLONS		
TI901	F33	30601-1	GAGE, TEMPERATURE REMOTE	LUBE OIL MANIFOLD		40°-240°F	
S380-2	E32	120623-6	SWITCH, PRESSURE	LOW LUBE OIL PRESSURE ALARM	45 PSI INCREASING 42 PSI DECREASING		
S397	C32	916156C1	SWITCH, Δ P	LUBE OIL FILTER HIGH Δ P ALARM	35 PSI INCREASING 27 PSI DECREASING		

Figure 4.4
Sample parts list that indexes to drawings. Corporate copyright. Used with permission.

24	23	22	21

SCHEMATIC FUNCTIONAL COMPONENT TABLE

REFER. DESIG.	DWG. ZONE	COMPONENT PART NO.	COMPONENT DESCRIPTION	FUNCTION	COMPONENT SETTING	SYSTEM NORMAL OR OPERATING DESIGN VALUE	NOTES
FCV901	C34	9116810	VALVE, THERMOSTATIC	CONTROL OF OIL FLOW TO OIL COOLER	VALVE IS CLOSED AGAINST SEAT AT 138 -142°F WITH LUBE OIL TEMPERATURE OF 150°F MINIMUM ΔP CRACKING PRESSURE IS 50 PSIG		
VCH902	C34	PART OF VMF921	VALVE, CHECK		2 PSIG CRACKING PRESSURE		
VCS902-1	B33	PART OF VMF921	VALVE, SPRING CHECK		2 PSIG CRACKING PRESSURE		
VCS902-2	B33	PART OF VMF921	VALVE, SPRING CHECK		2 PSIG CRACKING PRESSURE		
H901 AND B-78	E35		COOLER, AIR/OIL	140°F(60°C) AMBIENT TEMPERATURE LUBE OIL COOLER		FAN SPEED 1730 RPM NOMINAL WITH 2HP (1.5KW) 230/460 VAC 60HZ ELECTRIC MOTOR	
VH902-1	B33	9129360	VALVE, HAND	DUPLEX FUEL FILTER VENT	CLOSED		
VH902-2	B32	9129360	VALVE, HAND	DUPLEX FUEL FILTER VENT	CLOSED		
VH903	B32	9138700	VALVE, HAND	DUPLEX LUBE OIL FILTER CROSSOVER SHUTOFF VALVE			
VH904-1	A32	9138700	VALVE, HAND	DUPLEX LUBE OIL FILTER DRAIN VALVE			
VH904-2	A32	9138700	VALVE, HAND	DUPLEX LUBE OIL FILTER DRAIN VALVE			
VI901-1	H37	9153700	VALVE, NEEDLE	GAGE DAMPENING, LUBE OIL HEADER PRESSURE			
VMF921	D34	9192590C91	MODULE, LUBE OIL	LUBE OIL DISTRIBUTOR			
VR901	C34	PART OF VMF921	VALVE, RELIEF	LUBE OIL SYSTEM ULTIMATE RELIEF VALVE	130 PSIG CRACKING PRESSURE		
VR905	B34	PART OF VMF921	VALVE, RELIEF	PRE/POST LUBE RELIEF VALVE	10 PSIG CRACKING PRESSURE		
VT901	C33	PART OF VMF921	VALVE, TRANSFER	3-WAY SELECTOR VALVE FOR DUPLEX LUBE OIL FILTERS			

DRAWING NO.	REV A
41761-149201	SHEET 2

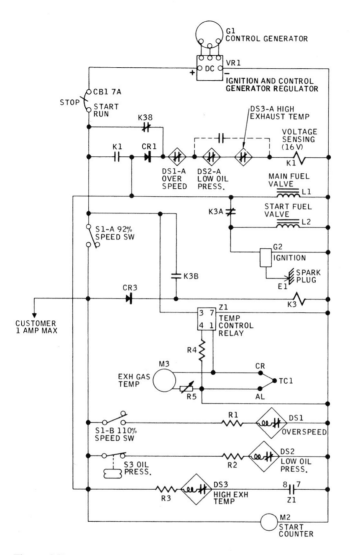

Figure 4.5
Electrical schematic. From *Selco Technical Publications Illustration Style Guide.* Corporate copyright, pseudonym used to preserve anonymity. Used with permission.

devices must attempt to capture the input that the designer filled in with his margin sketches, verbal communications, and physical presence on the shop floor along with other tacit knowledge necessary for repeated production of the same turbine engine package. The carrier of this information for any given engine package is the *build book,* known at Selco as a *traveler.*

Travelers and Build Books

In addition to informal sketches, occasional models, and fixed drawings, another type of visual information goes to the shop floor—a traveler. Travelers are various types of assembly information that accompany both in-house and vendor-supplied parts and travel along with the hardware. They are generated during the first assembly of the engine package. In the case of the Mark IV, photographs were taken during phases of assembly to capture how the engine should look during correct assembly.

As sketches and final drawings act as boundary objects in the conceptual phases of technological design, travelers act as boundary objects between designers and builders. While the final fixed drawing set emerges out of a shared cognitive process, the actual engine package emerges out of the process connected to travelers. The very purpose of travelers is to capture the process of building the engine package. They are in-house drawings that will not be seen by salespeople or by the customer but are kept in the shop for future production of subsequent packages. Unlike a layout, travelers refer to minute specifics; the difference between a layout drawing and the collection of travelers is like the difference between a map and a recipe.

The build book, made up of travelers, shows step-by-step assembly of each part. Carol, a mechanical engineer whose job is to put together build books, notes that "when they get done following this book, they'll have built the whole package." Build books at Selco contain visual depictions of various sorts, including crude hand-drawn sketches made by the mechanical engineer, photographs, and illustrations that are originally produced for customers' operation and maintenance manuals but are reworked to become integrated back into build books. Build-book drawings must contain more detail than the illustrations in the operations and maintenance manuals. According to the mechanical engineer,

They only go to a certain point in the operations and maintenance manual, and we have to surpass that. For instance, in, oh, this type of console we build, a seal oil console has a lot of switches and gauges. The O&M manual might show that

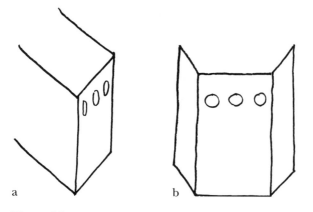

Figure 4.6
Author's rendition of sketches drawn by mechanical engineer. (a) Facsimile of computer-generated view of oil seal panel that machinists rejected. (b) Facsimile of the simple view that the mechanical engineer hand-sketched to replace the computer-generated version.

as switches, but they stop there. We have to show the mechanic out here that a tube goes from this switch to that gauge wall kit, and so we always have to add to the drawings.

Drawings are preferred for both build books and operation and maintenance manuals because their linear quality renders detail more articulately than the gray tones of photographs. Drawings made by company employees may be done using the CAD system or by hand. When asked if she used CAD drawings, Carol mentioned that workers on the shop floor had torn up some travelers as unreadable: they had come through the computer graphics system and were rendered in perspective but not the conventional 60 degree by 30 degree isometric view that is standard for illustrations (see figure 3.2). However, even that view is considered a pretty picture made for customers and not for production use—hence, the rejection by the shop-floor workers. Clear drawings follow the visual culture of two-dimensional rendition—plan, profile, and section views (see figure 4.3). The head mechanical engineer redrew the drawing by hand in the conventional format (figure 4.6b).

Art and Illustrations
Computer-graphics programs were used at Selco by drafters and detailers and not by senior design engineers. Drafters did create a computer model of the project as they put the designer's plans onto the CAD sys-

tem. Eventually they could produce a drawing that shows how the machine would look. But this drawing is a *wire model,* which means that all lines that represent all sides of the machine show at once as if the object were constructed of clear plastic or a wire grid. A senior designer noted that to give such a drawing to the shop floor would create more rather than less confusion, especially about small details. The excess lines on the wire model are eventually erased for the production of drawings for operation and maintenance manuals. But these drawings, called *art,* are not available during the first assembly of the machine. Art or illustrations are after-the-fact products that are produced after the first package is built. Any drawing that actually shows a view of how the completed package will look falls into this category. That is why the rough sketches provided for shop workers are so important. Illustrations eventually find their way back into production in parts manuals and build books after they have been produced for operation and maintenance manuals. Illustrative line drawings are labeled with component part names and numbers in operations and maintenance manuals to identify the location of various components on the engine package. Separate detail drawings give close-up and detailed views of the specific parts. Hence, instructions for such things as start up, lubrication points, and troubleshooting are accompanied by visual references. See figures 4.7 to 4.9 for examples of such art used in operation and maintenance manuals but not for design to production. Since these drawings take time and cannot begin until the first package is complete, labeled photographs often are substituted in the manuals for the first few packages delivered to customers. Illustrators also work from photographs—sometimes official ones taken by a professional and sometimes just color Polaroids they have taken themselves—to capture a precise image of a particular part before the actual machine is shipped off. This work was done by drafters and was being entered using the CAD systems.

Management saw the storage of components that regularly are used in illustration as an advantage. This is so to a limited extent. Since packages were being sold more on a custom level with little mass production of a common item, many changes were constantly made to the package. Using the computer-graphics system as a bookkeeping tool allowed for constant reproduction of illustrations with minor changes. However, if the changes were major, the cost effectiveness of taking the time to put them onto the computer-graphics system when they might not be used again was questionable since original drawing using the CAD system is much more time consuming than hand drawing. Management saw the

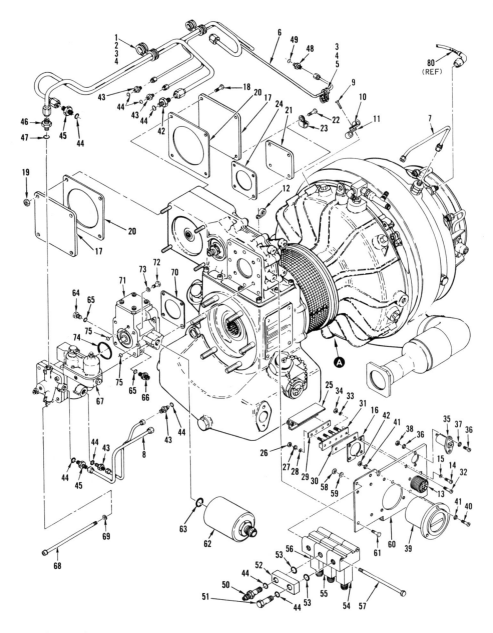

Figure 4.7
Illustrative art for operations and maintenance manuals. Strip breakdown method from *Selco Technical Publications Illustration Style Guide*. Corporate copyright, pseudonym used to preserve anonymity. Used with permission.

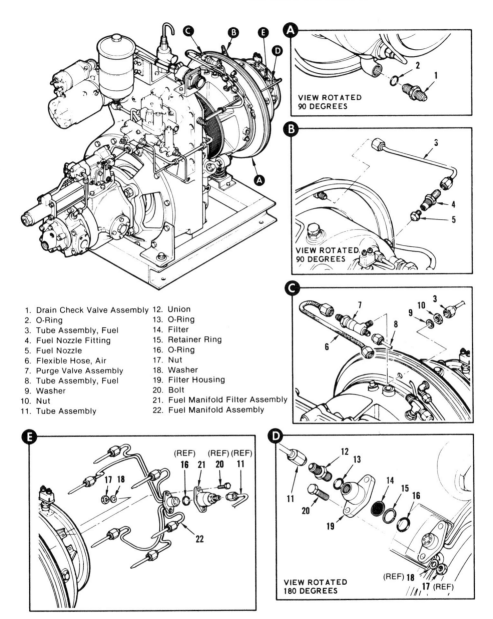

1. Drain Check Valve Assembly
2. O-Ring
3. Tube Assembly, Fuel
4. Fuel Nozzle Fitting
5. Fuel Nozzle
6. Flexible Hose, Air
7. Purge Valve Assembly
8. Tube Assembly, Fuel
9. Washer
10. Nut
11. Tube Assembly
12. Union
13. O-Ring
14. Filter
15. Retainer Ring
16. O-Ring
17. Nut
18. Washer
19. Filter Housing
20. Bolt
21. Fuel Manifold Filter Assembly
22. Fuel Manifold Assembly

Figure 4.8
Illustrative art for operations and maintenance manuals. Area detail breakdown method from *Selco Technical Publications Illustration Style Guide*. Corporate copyright, pseudonym used to preserve anonymity. Used with permission.

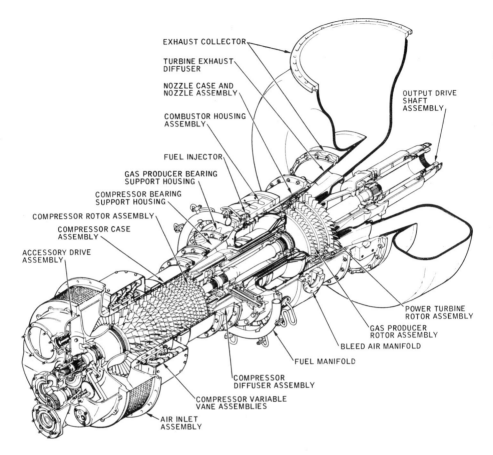

EXHAUST COLLECTOR

TURBINE EXHAUST
DIFFUSER

NOZZLE CASE AND
NOZZLE ASSEMBLY

COMBUSTOR HOUSING
ASSEMBLY

FUEL INJECTOR

GAS PRODUCER BEARING
SUPPORT HOUSING

COMPRESSOR BEARING
SUPPORT HOUSING

COMPRESSOR ROTOR ASSEMBLY

COMPRESSOR CASE
ASSEMBLY

ACCESSORY DRIVE
ASSEMBLY

OUTPUT DRIVE
SHAFT
ASSEMBLY

POWER TURBINE
ROTOR ASSEMBLY

GAS PRODUCER
ROTOR ASSEMBLY

BLEED AIR MANIFOLD

FUEL MANIFOLD

COMPRESSOR
DIFFUSER ASSEMBLY

COMPRESSOR VARIABLE
VANE ASSEMBLIES

AIR INLET
ASSEMBLY

Figure 4.9
Illustrative art for operations and maintenance manuals. Three-quarter isometric cutaway view from *Selco Technical Publications Illustration Style Guide.* Corporate copyright, pseudonym used to preserve anonymity. Used with permission.

aesthetics of computer-produced art as a selling point to convince customers that the company uses the latest technology. I have previously noted the parallel between the way perspective was used by illustrators to impress gentlemen entrepreneurs and the way computer graphics are employed for customer-oriented art. In both cases, the new illustration method gives the impression that design and production also use this latest technology, which was not the case here or historically (Baynes and Pugh 1981).

Drafters and illustrators were willing to learn the new technology and came in on their own time to practice while they were taking training classes because the skills appeared to offer the potential for future advancement. (This may be an illusion, as is discussed in chapter 5.) Unlike the design engineers, the illustrators and drafters (perhaps because of their lower status or because of less demanding responsibilities and deadlines) worked diligently to master the new techniques, contending with the slow building of an online parts library to customize the software for their needs. Some computer-graphics operators were also hired who already knew how to use the electronic graphics technology, but their lack of drafting and industry experience resulted in few of them being retained. The illustrators did experience problems, but they were relatively minor compared to the insufficient training compounded by pressing deadlines that the design engineers experienced. For example, they were not able to follow the exact arc of a piece of piping due to the limitations of the graphics system and therefore produced a slightly stylized version rather than photographic realism. This stylization makes all the illustrations nearly homogenous and gives the computer-graphics look that marks the work as distinct from hand drawings. One could argue that both aesthetically and for accuracy the look is actually inferior to hand drawing, but while the technology is still new the graphics aesthetic carries a certain status associated with the latest technological innovations. While many of the more obvious personal stylistics of hand-drawn work disappear when rendering with computer-graphics systems, some are purposefully retained (this is discussed in detail in chapter 5).

Another problem was isolation. Those who worked on terminals worked at cubicles in a separate, slightly darkened room, isolated from other illustrators who shared the same larger space and whose large drafting tables were gathering points for interactive discussion around drawings, the hub of activity. As mentioned above, the computer-aided illustrations for operation and maintenance manuals were intended for a customer-user audience, though they can filter back into production in build books. Yet the very fact that these pretty pictures are not aimed at production needs can potentially cause problems when they work back into production documents as illustrations.

The issues I have introduced here are a just few of the problems encountered when a new visual culture is introduced into the paper-world conventions of engineering, particularly at this field site. A larger sampling of issues and how they impact the work, workers, and interactions between inter- and intracompany groups and whole industries is explored in chapter 5. Here I continue to examine how the integration of the computer-generated drawings and other interlocking databases overloaded the official drawings, hampering their flexibility as boundary objects. This was the most significant problem that the introduction of the graphics system caused throughout Selco. In order to understand why employees at Selco become so intimidated by the databases, it is helpful, first, to take a closer look at the social context of the history of computer-assisted design at Selco.

Computer-Assisted Design at Selco

The Mark IV project was the first attempt by designers, management, and CAD-CAM operators to capture the design on the computer-graphics system early in the process rather than completely after the fact as merely a record-keeping operation. The package design manager describes Selco's usual practice:

We design things on paper, and then we hand it over to the CADs system. We use the CADs system as a record-keeping, and rather expensive, fast eraser.

He also notes that this is not regarded as the appropriate way to use a computer-graphics system:

We go backwards. Instead of building a picture and then taking the detail pieces off, we often end up going the other way around—which is wrong.

Actually, as is shown in chapter 5, many companies are using graphics systems in a number of ways that suit their needs. Such variation hardly makes those uses wrong, but the manager knows that Selco's approach is not the manner prescribed by the graphics industry. What is implied here is that to use the system only as a record-keeping device and a fast, expensive eraser will not justify its expensive price tag, which was sold with the argument that the system would cut costs by eliminating the need for costly redrafting between the numerous iterations from conception to production. Since design engineers were not using it in earlier phases, their manager could not claim such cost reductions as would match the claims of the sales hype. Note also his language use. Though CAD-CAM (computer-assisted design and computer–assisted manufacture) is a system designed to use computer assistance in both design and manufacture, this site was having enough problems implementing it in design alone. The *CAM* part of *CAD-CAM* was never mentioned and the grammatical use of the plural *CADs,* used widely at Selco, was singular to that site, producing frowns of nonrecognition or downright disfavor whenever I used it in other industrial contexts. Its use becomes almost symbolic of the supposedly anomalous manner in which the graphics system was being used. Such use was grounded in the manner in which the system was introduced to design engineers.

Selco management initially tried to interest its design engineers in using the new graphics system when it was first installed in the late 1970s. Classes were offered on a half-day basis: four hours of regular work and four hours of instruction for a period of two weeks. A senior designer stated, "A lot of us went through the course in CADs, but a crash two-week course in CADs could just leave you totally confused." He further pointed out that in comparing notes on course content with outside-trained CAD operators he found that the operators had spent "six months, eight hours a day" learning the same applications. Although a number of courses were offered to designers, not enough time was allowed for them to practice and play with the features of the equipment since they also had to keep up with their regular work.[5] Moreover, if designers did try to do their work using the graphics system, the initial bugs in the system in the early period resulted in the loss of hours of work. Engineers became disenchanted and infrequent users of the technology, thus making its use a time-consuming chore rather than an efficient technology. For experienced designers, hand drawing remained more time and cost effective. Management, encountering this resistance and having to meet deadlines, softened its promotion of CAD-CAM to a certain extent. It decided that implementation of computer graphics was going to be something that the next generation of upcoming design engineers would use. Since the company has been in a shrinking mode, however, the only new designers who work on CAD systems are actually serving as drafters and detailers, not conceptual designers.

While the subgoal of the Mark IV design team was to "do it right" this time by integrating the computer-graphics system early in the process, this did not include designing on the computer system. Rather, it meant digitalizing the design into the system before the package was built rather than after, so that the graphics capabilities could serve as more than merely an after-the-fact, record-keeping device. This extra step in copying lends itself to the potential for more frequent human error, such as transposed numbers and errors in dimensioning, but although some errors of this type did occur, these kinds of errors are common in design work and not particular to computer graphics.

A more significant problem at the Selco site was based in the formalization of the drawing in its role as a boundary object. The fluidity and flexibility that are part of the loose structure of boundary objects were paralyzed in this case because the whole system is computerized as interlocking databases. The problem was manifest in workers' reactions to this monolithic structure and their nonunderstanding of it. The hugeness and complexity of the interlocking systems intimidated people. Employees were afraid that if they adjusted information in their part of the system and made a small error such as the misplacement of a decimal point—a common error—that monumental consequences could result, such as the ordering of greatly excessive inventory, which could result in astronomical cost overruns to the company and the loss of their jobs. This fear was not unfounded given the many layoffs caused by shrinkage in the company and the industry. Moreover, employee attempts to use the methods that employees have always used to really get things done—informal means such as personal contacts, short cuts, and avoidance of official forms—compounded problems. The informal and the formal information structure and flow operating side by side meant communications breakdown for those depending on the integrated formal system to make sure everything was ready when the new design began to be assembled.

As an example of social impact on engineering, Latour (1987) cites Tracy Kidder's fictional *Soul of a New Machine* (1981). In that book, a computer design team's principle engineer takes apart a competing product and remarks that the design reflects the organizational problems of the company that built it. The best fiction is based on the realities of everyday life. In daily practice at Selco, the organizational politics that confined understanding to one's own department was reflected in the design of and interaction with the database. The system was treated like the man, Price, who designed it. As company folklore tells it, Price was overly particular in his applications procedures for computerizing data; he even wrote a book and gave outside lectures on computerization in industrial applications. A fastidious but not well-liked organizer, Price did not go far in his career at Selco. He was sent off on special assignment, officially isolated from the area where his ideas were employed. When the special assignment was over, he was told there were simply no openings back in the mainstream at that time. Company folklore attributes his demise to two factors—insufficient political clout and a design that was too meticulous. The status of the formal database is like that of Price—weak midmanagement, a position that managers can override and workers can circumvent. The methods used to do so are informal processes.

All this technological upgrading was taking place in the context of company cost-cutting strategies. The problem was further compounded by management's attempts to reduce paperwork. Selco used to use a form to initially notify departments about upcoming needs for support and materials and then modified the form later with specifics. In the new form, called a *preliminary change authority* (PCA), information was predefined from the database. Changes had to be made to the preexisting information from the database to initiate any action. There was no room for looseness and flexibility, such as keeping an excess of supply items in reserve for contingencies. The new form was supposed to follow an authorization signature routing so that all departments affected by a change were notified. While this could be very efficient, it was not. Few employees understood the interlocking relationship of the various departments throughout the company on the detail level of the database. This lack of understanding, coupled with their fear of making errors, led them to continue to use established workaround practices that circumvented official paperwork. While these practices had worked in the past, they no longer could function because the system had tightened through the use of the computerized database, cutting out the options to use informal networks to obtain needed supplies. When employees resorted to "hip shooting," in company jargon—skipping official paperwork so that forms such as the PCA are not routed to other departments informing them of upcoming needs—the system broke down.

Failure of CAD and the Database as Boundary Objects

Indeed, the failure to generate forms to communicate needs through the integrated database was exactly what happened in the case of the Mark IV. When the frame left the structures department, where it had been welded, and arrived on the shop floor, assembly of the package could not start. Standard parts—such as nuts, bolts, washers, and basic fittings used on all packages—were in short supply, as were more sophisticated parts such as gauges and valves specific to the package. Since interaction with the new online inventory control system had been circumvented, the parts had not been ordered.

Paradoxically, it was the computer-graphics department and the actual drawings they released, including schematics and parts lists, that employees came to use as an informal alternative to the intimidating database, though not without problematic consequences. The boundary-object drawing produced by the CAD operators allowed employees to see at least the project at hand in a kind of totality. Since the distributed information was perceived as coming from the CAD system operators group, that group become the liaison to production, though without the input of the design engineers. Hence, as corrections were made when common errors were detected on the shop floor, design engineers lost control of checking alterations. An engineer sketched out a hypothetical example. If the Structures Division, which does the welding of basic parts, finds an error in the official drawing—such as a transposed number resulting in an erroneous part number or the potential interference of one part with another because a dimension is too short or too long—it suggests a change, and the CAD system operator puts it back into the computer system without consulting the design engineers. Such hipshooting—instigating changes without proper distribution and analysis of the design content and economic impact—can have disastrous results. What looks like a simple change to someone who is not a design engineer may have significant consequences on the stress factors of one moving part interacting with another or may add unforeseen costs by requiring a more expensive part to make everything fit together.

The cost of keeping open a system designed to be closed is that a chunk falls out of the closed circle: that chunk is communication with design engineers. The drawings play the crucial role of standing in—at least on the project at hand—for the integration between the confusing details of many departments. Their flexibility is illustrated by the fact that employees have been able to substitute them for the closed database system to which they are indeed linked. A Selco project manager who had been working in liaison with various departments for eight years noted that probably twenty meetings a year address the proper administrative handling of design and production changes. He added that these meetings always become heated because participants became bogged down and lost in the details and fail to understand the total system of interaction. Thus, the difficulties that employees encounter in interacting with the database reflect the difficulties that departments have in interacting with one another. A boundary object that tries to encompass the totality of interlocking information in such a setting must either dictate the restructuring of the whole setting (a position advocated by many computer-graphics department managers who would necessarily advance to the level of a vice president) or they must maintain much more flexibility.

Conclusion

The new machine was designed to be modular. Modularization of the turbine package echoed and was constructed through the restructure of work distribution and actual workspace into modular cubicles and control devices of the company, while these structure and control devices were simultaneously necessitated by changes in the turbine engine and its market. Control devices—such as standards for how particular types of drawings should be made and for who may inspect and release them—delineate which yellow brick road should be followed. However, finding the road and traversing it to the Emerald City are not one and the same.

Those who make it to the end of the road must, as Cap says, keep putting out fires along the way because assignment to multiple projects means that they travel multiple roads at once. The success of the new engine is not achieved through the calculated moves and control structures that often hamper rather than aid the adventure down the yellow brick road but rather through the actions of designers, such as Cap, who constantly fill in gaps by chasing the fires to the shop floor, to the CAD room, and to meetings with customers.

When I arrived on the shop floor to observe as the prototype began to be assembled, the design engineers were a bit embarrassed. They assumed that I would expect each step to flow smoothly and perfectly in the linear process described by the myth of science into practice. It is the interactive knowledge—visual communication plus design engineers' enhancement of that visual knowledge with explanatory sketches—that gets the first project built despite the fact that the information is reconfigured at each phase from cocktail napkin sketch to model, early and late drafts of official layouts and schematics, build books, and finally art for customers. The sketches in the margins and the hand gestures on the shop floor fill in the gaps, while more formal approaches to control (such as the company's attempts to rationalize the design production process imposed on employee visual culture practices) have led to communication breakdown and a shop unprepared to assemble the new machine. Attempts to cover the tacit dimension, such as travelers and build books, work toward accomplishing the task after builders have already been through the task once and have added some tacit knowledge to the general stock of knowledge they acquired when working on similar projects. Nevertheless, companies officially claim that documentation has accomplished the task rationally and efficiently, while in practice sharing visual knowledge through sketches and travelers gets the job done, and computer-graphic drawings are recorded and corrected after the fact. Hence, in reopening the gray metal box of technological closure in the produced machine, closer examination of engineers' visual practices reveals a world in which interaction over paper with lots of sketching in the margins coconstructs a visual world as well as a new machine.

The successful production of the new design was symbolized in the sleek parallels of piping that envelop the bulky engine. As many as six to eight slender silver-colored tubes marched in a precision drill through crisp 90 degree turns around the Mark IV. This precision was a great contrast to the tangled mass of silver, black, and gray spaghetti that enveloped the Atlas V models, which was designed by the manager's favorite design team. The contrast between the two machines was the result of thoughtful design but also of the follow-through, patch-up communication practice of Cap, the senior designer and locally named guru of piping, who was on the shop floor every day hand sketching in margins and gesturing pipe bends to the workers who built the prototype. In contrast, the Atlas V designers used the over-the-wall approach and did not interact with the workers who built their designs except to send paperwork over to the shop floor—as if the information flow of the document without human interaction would accomplish the job. For the designers of the Mark IV, the yellow brick road included the final steps of ensuring that the tacit dimensions of their design received attention.

Here we find the essence of applying overly rationalized technical systems to the messy world of design production. The company attempted to control, dictate, and predict the behavior of humans, parts, machines, and information flow by setting up a highly complex, interlocking, rationalized system combining design drawings, inventory, paperwork, and large databases. Since work and the networks that support it are much messier than the designers of such systems realize, the yellow brick road does not map very well onto this superhighway. But in the meantime, the superhighway has made many of the roads in the old connected neighborhoods into dead ends, and when people try to use their old informal ways of doing things, they cannot get information to others who need it.

The design documents in this case become even more crucial because they have always served as a road map for everyone interacting with a project, even if designers have to chart the map during the trip. In attempting to connect the yellow brick road to the superhighway, even more messy patch-up work has to be done as designers attempt to get materials to the CAD drafters early enough so that the electronic version of the drawings will be done before the project moves from prototype to production. So there is much going back and forth as designers clarify communication with the drafters as well as with those on the shop floor.

In all these interactions, the drawings are the locus for the interactive co-construction of knowledge. However, the insertion of another element into their generation—the trip through the CAD center—causes confusion. People who use the drawings on the shop floor and depend on them for information when projects move to standard production typically have less contact with designers than in the building of the first production prototype. When common errors such as missing or conflicting dimensions need corrections, shop-floor representatives return to the source of the drawings. But now the bureaucratic paper flow, traced by dated signatures, shows that the source is the CAD center, not the designers' office, setting up another potential for error. While corrections that produce more errors are frustrating to all, the circumstances illustrate the centrality of the visual documents in the networks where decisions about machines are made. Indeed, even when closure occurs, and the new design goes into standard production, the documents too have produced a new offspring—the build book that accompanies the official drawings to describe the step-by-step process of assembling the machine. Nor does it stop there. More visual documents will accompany the new machine to its new home and new caretakers so they can get to know it through pretty pictures. Hence, whether at the initial conception, standard production, or users' end, those who wish to interact with the machine must do so through these conscription devices until they acquire the tacit knowledge through daily interaction to know the machine intimately.

This chapter has reviewed and illustrated how various visual representations serve to organize work and workers. The design team on this project had to overcome the unexpected turns and twists of the yellow brick road toward the production of their prototype and deal with a superhighway that created dead ends, but at least the team had collegial solidarity. In the second case study, discussed in the next chapter, conflict is a central issue. However, an important ally is the prototype itself, which as a member of the social network of its own production also had a political career that influenced the success of that network.

5

The Political Career of a Prototype: Development of a Precision Medical Instrument

This chapter focuses on the political career of the drawing and prototype pair—a linked series of visually oriented sets that resides in the borderland between research and development and production in industrial design engineering.[1] In this frontier space both consensus and conflict take place between groups and between individuals. The metamorphosis of design through interactions centered around the interlinked drawing and prototype pair involves the social construction of a piece of technology and the group process of the emerging design. Moreover, an examination of the process of design work reveals the strength of the visual—whether represented in three-dimensional plastic form or in two-dimensional drawings on paper—in organizing work, organizing knowledge, and recruiting and organizing resources, political support, and power.

Some authors suggest that the drawing process may be as important to design as the drawings themselves (Bly 1988; Tang and Leifer 1988; Bucciarelli 1994). The mundane interactions of actors, machines, and paper construct technological innovation. Throughout these interactions coordination and conflict take place over, on, and through visual representations because such representations are a component of the social organization of collective cognition and the locus for practice-situated and practice-generated knowledge.

The Study

The importance of sketches and drawings in engineering design as individual thinking devices as well as the locus for the shared cognition of team design work has been well established (Henderson 1991a, 1995a). The research presented in this chapter, undertaken with designers at a firm that produces high-precision medical optics, validates my earlier observations that visual representations serve as a social glue between individuals and groups and hence structure both worker networks and the work process itself. This study looks more closely at the processes involved in the interactive construction of sketches and drawings as con-scription devices that enlist participation and as boundary objects that facilitate multiple readings of the same material. The analysis examines how change in the social organization of participants and participation is connected to the metamorphosis of visual and plastic renditions of the design.

Latour (1986) notes how cascades of increasingly more informed engi-neering drawings facilitate the transition and transfer of knowledge from flat paper to machine. Here I look more closely at this ongoing transition from machine-to-paper to machine-to-paper-to-machine as it occurs in the workplace. The continuing series of paper and object pairs that I observed in the development of a new medical instrument served as (1) vehicles for organizing shared cognition in design work, (2) devices to capture threads of tacit knowledge, (3) control gates to determine who could or could not participate in manipulating design concepts, (4) emissaries to strengthen political networks, and (5) conversely, in the hands of detractors, evidence with which to attack design concepts.

A narrative account traces the construction and reconstruction of the design through the many lives of a drawing in its transmutations from flat paper to three-dimensional object and back again to paper and so on—a series of overlapping metamorphic dyads that serve to capture empirical knowledge in daily practice. Intertwined with this is what I call the *political career* of a prototype—an account sometimes articulated separately because writing is a linear exercise but intimately interlocked with the emerging design process. This *career narrative* shows how models and prototypes become, like Pasteur's bugs, actors in the political net-work of the company and garner support for the project through their role as conscription devices. As such they enlist supporters through inter-action and allow groups to impute differing interests to the same object. The political career of the prototype shares some characteristics with Goffman's concept of the moral career of a mental patient in that the role is constructed and reconstructed by competing institutional and individual interests but does not follow a single linear path or prescribed chronology. Just as access to the files that play a role in the construction of the mental patient are tightly controlled, so too are design drawings. Control issues—who is allowed to see drawings, manipulate paper or object actant pairs, or participate in their metamorphosis—are crucial in design development because visual representations may be read in a variety of manners, through filters of varying interests that can support or detract from a design project.

The use of computer graphics is another potential arena of conflict in the design process. Using ethnographic methods to look at the promise and the outcome of CAD-CAM applications, Downey (1992a) illustrates that adaptation to technology developments is taken for granted in the United States and that significant social changes in power relations are sanctioned without any public debate or even awareness on the part of those whose work will be most dramatically altered. This study looks at informal workplace resistance to potential change wrought by the implementation of computer graphics in engineering design. Such changes in technique can alter power dynamics and relations between workers as well as how work gets done. The research examines how engineers and designers managed to marginalize computer-aided design through benign neglect, leaving it to serve merely as a record-keeping tool for drafters rather than as a design tool for engineers. Through intentional nonaction designers kept their visual culture intact and kept the reorganization of visual representation by computer graphics at bay. This was done for at least two reasons: (1) to keep reorganization caused by the use of computer graphics from restructuring relationships between workers in ways that could hamper the critical flexibility of representations and (2) to keep such reorganization from influencing the alignment of various groups whose political support was needed for the new design to marshal sufficient in–house support to reach production.

Methods

Investigation of the social construction of technology necessarily leads to the examination of the daily processes involved in the work itself. The data presented here were gathered during my participation in the everyday activities of industrial engineers engaged in the actual process of designing a new piece of technology. Participant observation with the research and production teams centered around the weekly research and development design meetings, attendance at which varied from two to twenty persons, depending on the phase of the project. I also participated in after-the-meeting evaluation sessions with the project engineer and other designers and interviewed designers, drafters, technicians, engineers, consultants, marketing representatives, supervisors, managers, and the production engineer for the project as well as designers and managers not associated with the project. Additionally, I observed laboratory sessions in which various adjustments to or permutations of the new instrument were tested, and I observed drafting sessions in which versions of visual representations were adjusted, corrected, and improved.

Setting

The site of this case study, a company I refer to as MediVis, is known for its innovations in high-precision miniature lens technology. MediVis is financially healthy, a relatively young company, and a leader in the rising field of medical devices for eye surgeons. I gained access to the site through participation in a computer-graphics workshop where I met the supervisor of the drafting department that had recently implemented computer graphics. To attain access I signed a nondisclosure agreement stating that I would not reveal any proprietary technical information concerning the product except as approved by a representative of the company. For this reason, few illustrations accompany this research, and descriptions of certain aspects of the new technology are somewhat vague.

The MediVis plant is located in one of the so-called industrial parks that are filling in the canyons and orange groves of southern California. The site houses MediVis's entire miniature lens operations—from design through production, packaging, and marketing—in a sleek building and squeaky clean environment. Most employees use the modern and well-provisioned in-house cafeteria for lunch and coffee breaks. Casual socializing facilitates work-related exchanges of information along with the usual office banter in the wide corridors and elevators as groups go to and from breaks and meetings. Visitors check in with a receptionist at a large desk set facing a typically overcoordinated plush, wood, and chrome environment populated with lobby plants. As a weekly visitor to the research and development design meetings, I was allowed to check in, pick up a visitor's badge, and go to the upstairs engineering conference room without awaiting a personal escort.

The MediVis Injector

The project that I participated in was slightly unusual for the company. Rather than designing an innovation in the intraocular miniature lens, the MediVis hallmark, the designers on this project were involved in creating an innovative medical instrument for use by surgeons in the implantation of the tiny MediVis lens into the human eye. These high-grade lenses, less than one-quarter inch in diameter, are precision ground to individual prescriptions, just as a standard eyeglass lens would be. Eye surgeons use the minuscule intraocular lens to improve the sight of cataract patients. The goal of innovation in this arena of medical optics has been to produce technical enhancements to help surgeons reduce incision size so that fewer sutures are needed to close the wound, reducing trauma to the eye and recovery time of the patient. Toward this end innovations have included changes in the lens itself as well as changes in the surgical instruments used to implant it. For instance, an earlier innovative instrument folded the circular lens in half, thus requiring an incision smaller than the full diameter of the lens because the lens could be unfolded inside the eye. Eye surgeons have remained unsatisfied with such innovations because components have been difficult to manipulate and implantation techniques have continued to carry other potential trauma hazards for delicate eye tissues. As a leading producer of intraocular lenses, MediVis was keenly aware of the market demand for a simpler lens-insertion system within the ocular surgery community and was also aware that success in developing such a system could improve sales of their own intraocular lenses.

The designers' strategy for instrument innovation was to take advantage of the unique ability of a new lens material that compacts when compressed. If the size of the lens could be reduced during insertion, then the incision size could also be reduced. This would be accomplished by the new tool (called here the Optimed Inserter, a pseudonym) by drawing the lens into a stainless-steel tube by means of a retracting mylar paddle (figures 5.1 and 5.2). Only the (3.5 mm diameter) tube would be inserted through a small incision in the eye. The surgeon would release the lens into its appropriate location by controlled manipulation of a slide switch on the plastic body of the instrument holding the stainless steel tube (figures 5.2 and 5.3). Constructed of injection-molded, lightweight plastic, the whole affair would be about the size of a fountain pen and resemble a cross between a hypodermic syringe and an inexpensive ball point pen.

The metamorphosis of the new tool during the design process introduces another dimension to the form and function of visual representations as conscription devices and boundary objects. This dimension of design is the series of overlapping paper and object actant pairs—the

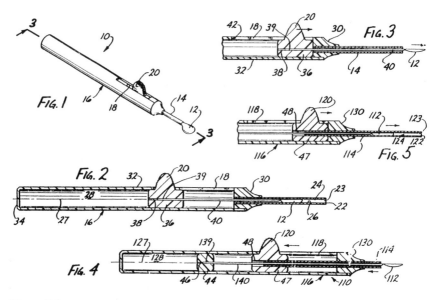

Figure 5.1
From U.S. Patent Document, Drawing Sheet 1: Figures 2 to 5 illustrate the working mechanism of the lens insertion apparatus. Corporate copyright. Used with permission.

various prototype objects in the sequence of drawn and redrawn representations of the design participants' shared knowledge. These prototypes, as interim concrete manifestations of the design, help designers to garner empirical knowledge that is incorporated into the final form and format of the designed object.

Such models and prototypes serve as boundary objects on at least two different levels. On one level, their flexibility is greater than representations of a design in drawings: because the design concepts are more accessible to a wider range of people, they are presented in a tangible three-dimensional format closer in form to the final product. This facilitates understanding by a broad viewership of nonexperts who may not be able to easily visualize a three-dimensional object from a two-dimensional drawing. On another level, for a narrower viewership of experts, the plastic versions can be viewed from the multiple perspectives of groups and individuals seeking empirical data. While plastic models are less flexible than sketches or drawings, which can be easily altered, the three-dimensional version provides, with relative ease, another level of information—that connected with touch, also known as *kinesthetic* or

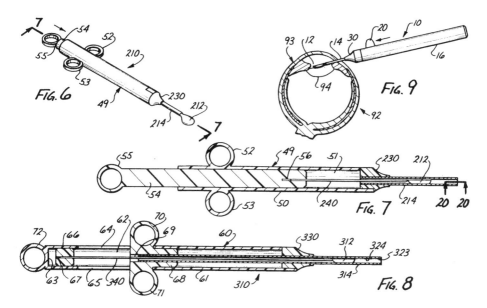

Figure 5.2
From U.S. Patent Document, Drawing Sheet 2: Figures 6 to 8 illustrate a conceptual variation in the configuration of holding device and triggering mechanism of the instrument, which were not prototyped. Figure 9 illustrates the application of the tool in eye surgery. Corporate copyright. Used with permission.

fingertip knowledge. This includes information such as how the instrument may be held and manipulated, how much resistance occurs between the moving parts, or how comfortable it feels in the hand during usage. Furthermore, feedback from the plastic iterations, as boundary objects tested by different interested contingents, is information from different viewpoints—that of a designer concerned with feasibility of function, that of a technician concerned with production possibilities, or that of a physician who will ultimately use the tool in surgery and is presumably concerned with the impact on the health of the patient in addition to his own reputation.

The simplest way to understand how a series of metamorphic paper and object pairs works in the design context is to examine the many lives or various phases, purposes, and uses of a drawing on its way to production by means of a narrative account. Of course, simultaneously the political career of the prototype pair is being developed. For clarity, however, the following account first traces the development of the multiple iterations and tentative models of the Optimed Inserter.

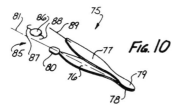

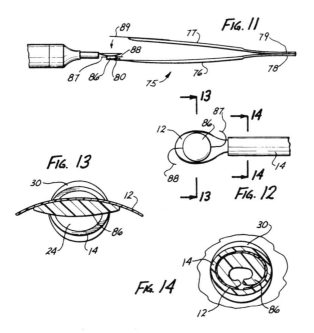

Figure 5.3
From U.S. Patent Document, Drawing Set 3: Figures 10, 11, and 12 illustrate lens location for loading into the instrument. Figures 13 and 14 show lens configuration inside and outside the injection tube. Corporate copyright. Used with permission.

Construction and Reconstructions

The Many Lives of a Drawing

The first documentation of ideas in the form of drawings for the Optimed Inserter project emerged from sketching conversations between Sam, a technician, Tom, a drafter, and Dale, the project engineer. Technological design seldom starts from thin air. A small (about two-inch) medical syringe was a starting point for developing the functional capacity of the Inserter, while a nine-foot kayak, a hobby project of the original drafter assigned to the Optimed project, was the inspiration for the physical form of the first design and prototype pair. The designers saw this form as ergonomic—a shape that would aptly fit the contours of the surgeon's palm. Other employees referred to this design as the fish. The overall appearance of this first model resembled the toy musical instrument called a kazoo: the slide switch projected on both sides of a flattened, oval tube that broadened just beyond its center point. While the unusual physical shape of the first iteration of the Optimed Inserter did not survive the first round of feedback from the eye surgeons, many of its conceptual elements and interior functional components were retained and refined in later versions.

The importance of the first set of drawings, prior to the first plastic model of the project, is expressed by Frank, a consultant in mold-injection techniques who provided manufacturing input during the development process:

They talk about—they want to make a [new tool]. We use that as an example, and everybody's got an idea that it's just this vision out in the air. . . . The engineer or the designer or whoever puts it down on paper, he then becomes the first person to get second guessed because he's taken it from an idea and made it into a visible, concrete drawing or idea, whereas before, it was, "Hey, we think we can do this" or "We think it will do that." Now he's put it down on paper, and he has now made it a concrete, visible entity, at least to start. And [then] somebody says, "I want this done" or "I think we should make it this way" or "Maybe it'd be good that color."

Frank is pointing out that as soon as the first sketches are shown to another pair of eyes, suggestions for changes start. Ideas expressed only verbally can be forgotten, but once they are made specific through concrete depiction, they gain in stature. This illustrates the conscriptive quality of the drawing itself in that for any design suggestion to carry any weight at all it must make the first transition from, in Frank's terms, a "vision out in the air" to "a concrete, visible entity." The tension created by this transition is that it results in the creation of a tangible boundary object, which can now be interpreted from a variety of view points—or "second guessed," as Frank says. For this reason, those who work in design try to keep access to drawings limited to those whose input and motives they trust. An example is Frank's participation. He joined the design team as a manufacturing consultant to the Research and Development Department. Though thoroughly knowledgeable about molding technology from experience in a previous job with a pen

manufacturing company, Frank was not a member of the Medivis Manufacturing Department and did not represent its interests. Rather, he was
employed by the Research and Development Department directly. The
Manufacturing Department, however, was not consulted in these early
stages. This was intentional, for the reasons mentioned above. The continued absence of Manufacturing from the design process, even though
manufacturing input was provided by an in-house consultant who was
not attached to the actual Manufacturing Department, ultimately introduced political problems that influenced the technical problems and
vice-versa.

Initially, Dale, the project engineer, and Sam, a highly regarded technician, came up with the basic design for the first model. Frank, the
manufacturing consultant, provided input on the viability of injection
molding the instrument, and the drafter provided specification input.
The drawings were then sent out to a model maker to create a first approximation of the instrument. Latour (1986) has discussed the layers
of information captured in maps and engineering drawings and noted
their facilitation of the transfer and translation of knowledge from flat
paper to machine. This is an imprecise and messy transfer that moves
back and forth between machine and paper, each enhancing the next
rendition of the other. The continuing series of paper and object pairs
that I observed in the development of the Optimed Inserter illustrates
not only the interactive nature of shared cognition in design development but also the role played by prototypes in capturing threads of tacit
knowledge and as devices to strengthen influence in political networks.
The first sketches and drawings of the Optimed Inserter helped organize
the interaction of the close-knit team members who originally worked
on the design. They also clarified team ideas for the more general viewership of management and other departments. The subsequent model
served several purposes, not all directly related to form and function.
As the project engineer stated:

[pointing to a drawing] This was basically our first rendition, and we worked
from *this* particular drawing. And what we did was we said, "Well, OK. We were
coming up with two different designs." We wanted a sliding tube and a sliding
paddle for the slide switch. . . . And we wanted to see what the effect of that
was. In the terms of operation, it became very clear, once you tried out a prototype, which was the best method. The only thing was that it was clear to some
people, but it wasn't clear to others. So it had to be shown very clearly—by practical [means]. And it's always best to communicate something by building a prototype and then showing everyone involved. And if it can be clearly stated, it's best
to communicate ahead of time so you don't have to build a prototype to do it.

If certain things aren't clearly understood, then you have to build a prototype to test it out. And it's just a matter of how far the general understanding can be brought up through just purely verbal communication. Then there's a limit. Then there's gonna be some people's thoughts involved and that's gonna hold everything back. So what needs to be done then is to either draw it on a piece of paper, or . . . [if] it doesn't help, it needs to be built into a prototype.

He further elaborates on the benefits of prototypes over drawings alone:

It [a drawing] helps in terms of, "Oh, it looks pretty" or "How does this [work]?" or "I think my finger could work really well on this." But until you actually get your hands on it, it's still not even close. There's a big, big difference in drawing it and making it. That's really where the main change comes over— when you get the prototype. . . . I would say you cannot make anything unless you make prototypes, um, and understand it and worry out the bugs.

The project engineer is pointing out several important functions of the prototype. Many of these functions overlap in serving to enhance both practical design consideration data and political network-building influence. The design category functions include (1) empirical information needed to decide between two approaches to the design, (2) empirical evidence to substantiate that the functional concepts of the sliding paddle, which was retracted and extended via the slide switch, and the slide switch itself actually worked, (3) elicitation of tacit knowledge in the form of feedback on how the instrument felt to people—its ease of use, fit in the hand, appropriateness of size, and so on. A fourth function, equally related to the political category, was to illustrate the successful function of the design concept to the medical community and company networks needed to maintain smooth working conditions so that individuals and departments would give the project timely financial and personal attention and support. That garnering such support is as important as the actual functions tested by the prototype is illustrated by the problems that occurred late in the design cycle despite attempts to avoid the usual divisiveness between the Research and Development Department and the Manufacturing Department. The project engineer mentions the importance of prototypes in influencing supporters, first emphasizing the importance of visual renditions versus verbal information for clarity. But in his statement that "there's a limit. Then there's gonna be some people's thoughts involved and that's gonna hold everything back," he alludes to more than noncomprehension. He is suggesting different interpretations. Hence, the role of the prototype is to generate clarity and to conscript support through elicitation of visual and kinesthetic information to be incorporated into drawings.

The project engineer's comment that "you cannot make anything unless you make prototypes . . . and understand it and worry out the bugs" references Law's (1992) concept of the economy of prototypes. Law notes that while designers are expert in visualizing three-dimensional structures into two-dimensional drawings, "it becomes increasingly difficult to be sure that what appears to be possible or desirable on paper is indeed so in solid form" (417). His study documents the series of increasingly costly and increasingly less flexible representations of the Olympus 320 engine, culminating in a fully operational prototype. The early engine models, like the prototypes in this study, were used to elicit feedback, allowing revisions to drawings that would be cheaper than subsequent changes once metal had been cut for the aircraft engine or molds had been cast for the medical instrument. The economy of prototypes facilitates modeling both the multiplicity of parts and the interdependence of parts during performance. Law emphasizes the absence of any strict chronology of prototypes or tests, noting that the quality of the testing of a range of engines was nonlinear and that engineers expected failures, which would lead to modifications, more tests, and further modifications.

Similarly, at MediVis, in order to test both the viability of the medical instrument design and the willingness of physicians to accept and use it, the project engineer took the model generated from the first set of drawings to an ophthalmological products exhibition and showed it to several eye surgeons. They were very enthusiastic about the functional concept of the instrument but critical of the actual design. They disliked the oval tube and the bulging ergonomic shape. The most prevalent feedback was almost unanimous preference for a simple round tube and a single actuator accessible from one side of the tube only.

The Metamorphic Dyad: A Paper and Object Actant Pair

The importance of the metamorphic paper and object pair cannot be understood in terms of the prototype acting alone. As Law points out for turbine engines and metal, design concepts become literally fixed and remain primarily unmalleable when molded into plastic (with the minor exception of the ability to shave some working parts to make them slightly smaller). Design participants must return to paper representations to access the flexibility of sketches and manipulate the design even if changes are not of an immediate visual nature.[2] Often designers will pass a prototype back and forth between them while they discuss and point to portions of the prototype that need to be altered and someone

simultaneously sketches the changes. The sketch may also be collaboratively constructed when individuals draw over portions to clarify them
(see figure 4.4). The prototype is both the source of the old knowledge
and a vehicle for generating new knowledge, while the sketch serves to
record the changes as a form of visual note taking. The return to the
paper rendition in the presence of the model or prototype allows designers to incorporate the new knowledge that has been generated from
interaction with the literal manifestation of design ideas into the next
iteration of the design. The plastic rendition is an intermediary source
of tacit knowledge elicited between the paper design rendition that preceded it and the one that will follow it. The new paper representation of
the design is sometimes merely an altered version of the earlier drawing,
highlighted to show relatively minor corrections and then cleaned up
by the drafter. However, often the new representation is so altered from
the original drawing that a completely new drawing must be created.
This was the case for the design team for the Optimed Inserter.

Having gained feedback from the initial model shown to physicians,
the design team was determined to generate as much empirical knowledge as possible before producing a completely new set of drawings.
With the support of other design team members, the team technician
created what is known in industry jargon as a *quick and dirty prototype* to
elicit more empirical knowledge and company support for the project.
A quick and dirty prototype is one that is not created from precise design
specifications but rather is constructed in the workshop from preexisting
materials and assembled to approximate the ultimate form and function
of the actual product. As the technical consultant, Frank, described it:

Even before the drawings were generated . . . or at the same time, Sam was able
to get some [prototypes] made because it was so simple: it was three-eighths
tube with a hole in it. . . . He had them sent out but he did it real quick and
dirty because he was able to go to somebody and say, "Hey, I need a three-
eighths diameter tube with a hole in it."

The plastic three-eighths diameter tube used for the model was easy to
obtain because it is widely used in plumbing and irrigation systems. The
plastic that eventually was used in the molded instrument could stand
up to the federal sterilization standards for medical instruments. The
quick and dirty prototype at MediVis is similar to a model in that the
prototype and the actual product resemble one another in design and
appearance while the methods and materials of their creation differ.
The scale is one to one: it is a fully functional, working model and not

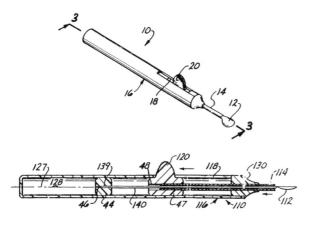

Figure 5.4
From U.S. Patent Document, Cover Representation Showing Early Conception of "Small Incision Intraocular Lens Insertion Apparatus." Corporate copyright. Used with permission.

a mere shell illustrating exterior design. It is this initial quick and dirty prototype that is illustrated on the cover and first drawing set of the patent document (figures 5.1 and 5.4). Other configurations were submitted as well (figures 5.2 and 5.5) to ensure broad patent coverage, though they were not built as prototypes.

The designers continued to tinker with the lathe–cut, quick and dirty prototypes, testing various mylar thicknesses and paddle sizes as they simultaneously finalized the new set of drawings for the injection-molded, final version of the instrument. The production engineer described the tinkering, the interactive nature of design work, as well as designers' practice of temporarily abandoning a design to return to the task with a fresh vision:

Sam works mostly on the instrument, and what happens is I usually just go in his office, test it out, and say, "Oops, this doesn't work right," and just change it. I'm basically the person who comes in and says, "Ah, this isn't working right," and then I say, "Let's change this and see if this will fit and get rid of that problem." . . . Or I bring them into my office and play with the instruments. That's when things get accomplished . . . when I'm grinding at the bit and that's just the next thing on the list. One of the critical things to know is that if you keep working on the thing all the time it doesn't get any better. You know, you can work superhard on it, but it just doesn't improve. Going every time with a fresh approach, with a fresh look, and not being afraid to change it is just as important. And always realizing it's never perfect and that it can always be improved, and looking for that: that's how it improves; that's what's important.

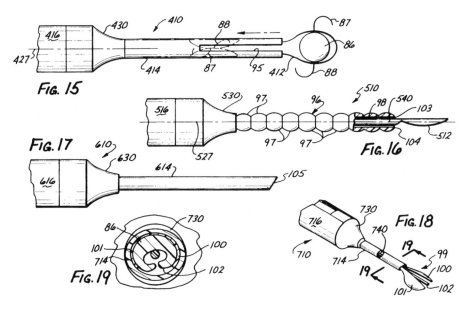

Figure 5.5
From U.S. Patent Document, Drawing Set 4: Figures 15 to 18 illustrate conceptual variations in paddle and tube configurations. Figure 15 represents a slot mechanism to hold the paddle, not prototyped. Figure 16 represents a variation in shaft design, not prototyped. Figure 17 illustrates a beveled tube end that was implemented late in the design process. Figures 18 and 19 represent a two-piece paddle, not prototyped. Corporate copyright. Used with permission.

Tinkering in these sessions, before and after design meetings, and in lab sessions appeared to resemble play. Indeed the product engineer uses the term *play* for what he does with the instruments. This play involves making the instrument perform over and over while closely observing how well it functions, viewing it from different angles, and holding it in different positions. Such actions facilitate seeing it with the fresh look the project engineer describes as crucial to innovative work. His emphasis on "play" and "keeping a fresh look" versus "working all the time" and "working superhard" centers on the felt and seen information that interaction with the tool generates and contrasts with notions of purely rational and objective approaches. Design team members join in such play as they pass the tool back and forth among themselves. Such interaction, centered on the prototype, again illustrates the plastic object's role as conscription device and boundary object. The prototype is an actant in the play as well as a centering device for concentration of visual, fingertip, and other tacit knowledge shared by participants in the interaction. This knowledge becomes articulated into a verbal format that relies heavily on visual reference. Communications take forms such as "this little piece, here," or "where it protrudes, there"—articulations similar to those Knorr Cetina (1990) reports in scientific practice. She terms this *optical induction* in that the visual portions of the operations are linked to the talk by physical gesture and place holders in the conversation. In this engineering design setting, play helps elicit and construct individual observations and knowledge. Its communal nature builds both on individual contributions and on interplays between contributions so that the outcome of even one such encounter is a rich ground of interactively constructed and distributed knowledge. This new knowledge is then collected from the interaction with the prototype through note taking in sketches and remarking of existing drawings.

The second plastic iteration of the design in the form of the quick and dirty prototype helped designers finalize the form of functional parts and the exact dimensions overall and capture collective changes in sketches and redrafts of previous drawings. The early drawing set for the ergonomic version of the design had been entered on the computer-graphics system. However, since the design changes were substantial, the early drawings were of no use, and a completely new set depicting the round-tube, one-actuator version of the Optimed Inserter was generated on the computer system. As the initial ergonomic-shaped version faded into oblivion, the recognition of team input to it also faded. As one team member put it, almost everyone in the company credited Sam the technician for the design of the Optimed Inserter, probably because he was the person who was able to share an early prototype. However, the design was truly a team effort that included input from the technician, the product engineer, the manufacturing consultant, the drafter, and the physicians. Sam built the quick and dirty prototypes, but the team designed them through sketching and discussion sessions of initial ideas, followed by sketches, often-altered drafts, and clean drawings and finally produced the first ergonomic model. More sketching and talking among designers and between designers and doctors and between designers and the manufacturing consultant preceded the quick and dirty prototypes. These prototypes, in turn, facilitated tinkering and feedback, more tinkering and more sketching in the messy, nonlinear process moving toward a final, well-designed product.

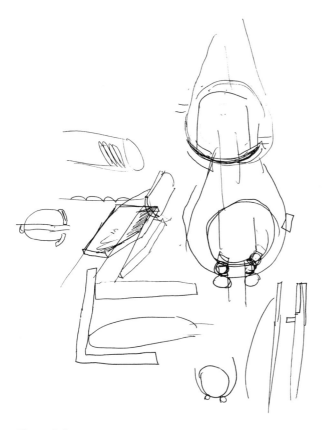

Figure 5.6
Talking sketch drawn by two engineers, passing one pen back and forth. Discussion concerned detail of forceps design. Corporate copyright. Used with permission.

Figure 5.6 illustrates an example of the "talking sketches" that occur throughout the design process, particularly between prototype iterations. This one was produced by two individuals talking and drawing using one pen. They were working out details to improve the forceps for loading the lens. The capture of tacit knowledge gained through such collaborative interactions between individuals, the prototype, and pen and paper in the act of sketching, however, is the same.

Refinements continued to be incorporated throughout the interactive design process. Dale, the project engineer describes how empirical information influenced size refinements on the paddle:

We first started off with a paddle that was way too big. And, ah, by cutting out paddles and putting them next to fake eyes[3] you can see how big the paddles were and realize how much smaller they must be for them to be a reasonable design. So we decreased the size. We decreased it so much we're now going up a little. We're adding maybe half a millimeter. Nothing significant. . . . What we do is just test out what the functionality is and then make it smaller. And you make it smaller. And you make it smaller. 'Cause in this case smaller is better. . . . And then you go up little bit and you don't have any quality control problems. Um, you want to have enough fat in there that if you used it you wouldn't hurt anything.

Having solved most of the design problems to their satisfaction, and having tested them to the extent possible with the quick and dirty prototypes, the designers sought an outside vendor to mold the instrument. Though some design team members suggested manufacturing the tool in-house, other more influential members of the Research and Development Department insisted, for historical reasons, that an outside vendor produce the initial run of instruments. The design team itself chose the vendor. One interesting factor in its selection process was that the vendor was required to have "high tech" production capabilities, even though the job required no special new materials, innovative techniques, or particularly original conceptual approaches. (I take up the symbolic aspects of the "high-tech" designation in chapter 8.) Another factor that the designers readily admitted influenced their decision was that Fabrimed, the vendor chosen, was the only company of the three submitting bids that submitted a new and altered version of the drawings. These new drawings illustrated minor changes that would improve the instrument and make molding easier for the fabricators. The appreciation for Fabrimed's attention to levels of detail is illustrated in the manufacturing consultant's account:

Fabrimed took it and says, "Here's what we propose." They came back with a lot of those things [in drawings]. They sent back an assembly [drawing]. . . . And then from that assembly we sat down and talked about why they wanted to do this, and they explained it. . . . So then when Fabrimed got it, they came back with a proposal. . . . The design did not change; just the way they assembled it changed. They used a refinement of the design for manufacturing. Once Fabrimed got a hold of the, um, the instrument, one of the changes they did, they made the body two pieces for ease of molding. The, ah, end cap, which holds the stainless steel tube, they put an under—what we call an "under cut" or "capture ring"—to capture the front of the body piece, and you can see that in the drawing, I'm sure. . . . They were afraid that unless they captured that it would spread open. They couldn't guarantee that the slot would be the same width all the way down, so they put a ring on the foot that captures the body and prevents it from opening up.

Fabrimed's drawings were assembly drawings that show how things go together in a general overview but are not sufficiently specific for actual shop floor production, such as machine grinding or molding. The drawings that MediVis later received from Fabrimed were very detailed. Frank describes set 2, from Fabrimed, contrasting it to set 1, the preceding set that MediVis had generated and sent to the vendor as design source documents:

If you look at these [set 1] and you look at these [set 2], there's a distinct difference. I mean they're like night and day. These [set 1] have a length and a diameter. Those [set 2] have length and diameter, draft dimensions, wall thickness. . . . Highly detailed prints, highly detailed prints from Fabrimed.[4]

These "highly detailed prints" from the designated "high-tech" vendor were all hand rendered as opposed to computer generated. Every drawing—from the initial proposal assembly drawing, which was included in the company's bid for the production contract, through the later, highly detailed production drawings used to generated the mold drawings—was drawn by hand. At MediVis the new drawings from Fabrimed were then regenerated into the computer-graphics system. The old computer-drafted drawing set, referred to above as set 1, was not updated but rather abandoned, and the new, hand-rendered drawings (set 2) were copied by a drafter into the computerized system. As Frank stated: "They simply copied Fabrimed's drawings. . . . It was just a direct copy of the text prints, just to put it on the computer." So set 3 of the drawings was generated as an exact copy of set 2 with the exception that set 2 was hand drawn and set 3 was digitized into the computer-graphics system. Since the vendor, Fabrimed, was the originator of these drawings and would have to be involved in any changes, set 3 was not a working set of drawings that would be altered for slight finishing adjustments. Rather, it was a file copy, much like a photocopy. Hence, set 2 was really the final set of working drawings, the set over which discussions and negotiations between the designers and the fabricators took place, the set that was redlined and corrected and regenerated by hand drafting as the final decisions were solidified. The mold drawings were then generated from the final version of these drawings.

Fabrimed alone handled the next series of paper and object actant pairs as its mold designers began to work from their redrawing of the Optimed Inserter. These mold drawings, along with the tooling for the molds and the molds themselves, remained at the vendor site where they were inspected and passed by representatives of the Optimed design team but not by representatives of the Manufacturing Department at MediVis—a step intentionally avoided by the design team but one that would come back to haunt them.

The Political Career of a Prototype

The completed molds issued another set of prototypes, referred to as *pre-production prototypes*. These differed from the quick and dirty prototypes and the first approximation of the design model in that they were produced from the actual materials and processes from which the final product would issue. Their explicitly acknowledged industrial purpose was to work out final bugs and refinements. The product engineer described pre-production prototypes:

Production prototyping is basically making up 500 or a 1,000 of these instruments, giving them to different doctors for their use, or testing and getting feedback from them, making minor modifications to the instrument. And then we go into full-scale production. What we try to do ahead of time is work out any bugs that may occur, so nothing like that happens.

Asked how designers prepare for bugs ahead of time, he gives the following example:

Well, let's say we've got different size lenses. You take the biggest lens you have, and you take the smallest lens you have—those are the extremes—and you see what effect it has on the lens and the loops and all that. We take, um, different ways of putting the paddle on the lens and see how much variability: if I slip it this way, if I put it this way, what's going to happen? You try different tip configurations. You try different polishing things, and you have to understand what are the important factors for providing a good instrument. What are those factors that are essential? . . . And those would be something that would be polished well—has to be a critical factor. And you have to define what "polishing well" means. What type of polish, what edge finish we'll get. What radius you expect it. . . .

   You have to go and specify everything. If you don't get into the details, they come back and haunt you. . . . Now maybe the specification of the length is not that critical. We can make it a little longer or shorter—a quarter of inch, a half inch. It's up to the designer to come up with it. These things aren't critical. It doesn't make any difference. . . . You just say, "It's close enough." But other things, there are right and wrong answers. You have to make sure you're in that box of right answers all the time. . . . Now we're looking at going into prototyping, and after prototyping, then we'll go into full production. All prototyping is sort of proving that you haven't been foolish.

At this point in the narrative account, the drawing has been through the major trajectory of its metamorphosis and is fairly stable. However, its role as a conscription device has passed to the small contingent of new pre-production prototype samples. These prototypes enter the competitive world of corporate capitalism in a role that is a mix of political organizer, heroic survivor of innumerable trials, and recruiting sergeant. These recruits must go through series after series of tests and trials to prove they are capable of doing their job and capable of meeting medical and federal standards for use in medical practice. At the same time they must convince advocates and adversaries alike to support their candidacy as a profitable product. The project engineer sums all this up: "all prototyping is sort of proving that you haven't been foolish."

In design meetings at MediVis as well as other sites, the conscriptive quality of the visual representations was so strong that conversations would falter, and someone would often leave meetings to fetch forgotten drawings so members could proceed with the discussion. This conscriptive quality extends to the prototype as part of the paper and object actant pair. The prototypes were brought to all design meetings by Dale, the project engineer, or Sam, the technician. When I, as a newcomer, attended my first design meeting, I was handed one of the prototypes and encouraged to play with it. This was sufficiently early in the design stage so that only a couple of the quick and dirty prototypes and the original ergonomic model were available. Dale and Sam were still testing various sizes of tubes and paddles. They asked me how comfortable the tool felt and whether my finger felt right on the slide switch. They watched to see how I held the instrument without instruction, and queried me on how smoothly I thought the slide switch moved. They then handed me a minuscule lens to try in the instrument and showed me the correct way to load it. I pulled back on the slide switch, and the lens disappeared into the stainless steel tube that protruded like a hypodermic needle from the instrument. The lens reemerged in perfect condition when I pushed the switch forward—a performance that had impressed optical surgeons and would later impress corporate lawyers sufficiently to gain their support in an attempt to obtain a high–tech designation for the new tool.

Newcomers from other departments and visitors to design meetings were provided a similar experience. All were encouraged to play with the instrument and their feedback was solicited. These actions actively engaged them in the product. The prototype was doing its work. Not only was it proving that the designers "had not been foolish," it actively engaged newcomers through their inquisitive and successful manipulation of it, while the designers reinforced the experience by soliciting feedback. The success of this engagement was illustrated when visitors asked for samples to take back to their own departments to show to others, thus furthering the support network. The prototypes were doing their recruiting job: people wanted to play with them, demonstrate them to others and show they were in on the new product.

As the basics of the design for the product design solidified, the original design team refocused on preparations for the incoming flux of prototypes that Fabrimed was tooling up to produce. Simultaneously, the nature of the design meetings began to change from the easy camaraderie of the small group as representatives of departments other than research and development joined the team to facilitate the product's change from experimental to production status. When only one or two new members attended the meetings, their informal introduction to the prototype in the manner described above appeared successful. However, as new team members returned to their own departments, where other interests might take higher priorities than the development of the Optimed Inserter, problems developed in the transition from design to product. Tensions underlying these problems can be understood by viewing the prototype actants in their role as boundary objects. To provide a better view of the multiple perspectives applied to the prototypes, I recount the emerging political perspectives and structures as they evolved in the project.

Joan Fujimura (1988) introduced the concept of a scientific bandwagon that comes into existence when a large number of people, laboratories, and organizations commit their resources to one approach to a problem. In an industrial setting, support networks are a somewhat similar microversion of the same phenomenon. For a design to come to fruition, powerful persons and groups within the company must jump on the bandwagon or, to use the capitalistic jargon of industry, they must buy in to the project. Buying in is apt in that it connotes the willingness to advance capital in the form of tangible expenses and personnel time to invest in the success of the project. Despite a powerful network of supporters, certain problems during the project's production phase stemmed from crucial individuals and groups that had not bought into the project. This resulted in two levels of friction that slowed down the project. Some individuals gave the project low priority and continued to practice business as usual, which meant that papers might sit on their desk for weeks while they paid attention to projects they regarded as more important. A source of more overt friction came from certain department representatives who had agreed to allocate time, space, and personnel to the project at one point and then later refused because of what they perceived to be the best interests of their own department and their own career within the company. Since these priorities and alliances changed over time, a narrative format is resumed here for clarity.

During the generative phases of development the project was under normative Research and Development Department auspices, funded through the department's general budget. At this point the project designers were essentially left alone and given creative space and time to generate ideas. Once the designers concluded that the design concept was viable, they took it to the ophthalmological instruments exhibition for validation by its potential market—physicians. Even the initial ergonomic iteration garnered doctors' enthusiasm for the concept and bolstered sufficient confidence in Research and Development to earn the project its own budget. Subsequently, the busy little quick and dirty prototypes did their job and did it well. They traveled outside design meetings and gained some powerful allies in upper management, who read them from the perspectives of marketing and company profits. This orientation led to a decision to market the new instrument in tandem with the company's lenses: only physicians who used the MediVis intraocular lens would be able to take advantage of the new surgical tool. Support for the project came from the level of company and division presidents, vice presidents, legal advisors, top marketing managers, and top research and development managers. As the excitement over the new product mounted and it looked like it would be especially successful, other departments that had not been involved in the product wanted to jump on the bandwagon. In particular, the Manufacturing Department, which had not participated in the earlier design sessions to develop the product and was not participating in its production, was invited to participate, late in the cycle, as a sort of peace offering for its earlier exclusion.

As mentioned, the design team—on advice from certain members and management who knew about past friction between the Manufacturing Department and Research and Development—had intentionally decided to send the production contract for the Optimed Inserter outside the company. Past experience with bureaucratic, procedural, and personal ego clashes were the reasons given by the design team. This kind of friction between Research and Development and Manufacturing is fairly common throughout industry. Mike Cooley (1987) points out that this conflict is often due to the separation of hand and mind, which leads to abstract, mathematical design work that can be a disaster in the shop. The Research and Development team tried to side-step this classic friction by having their new product produced outside the company. When the Manufacturing Department wanted to jump on the bandwagon, it was given the packaging of the product, a fairly involved, triple process for a medical instrument that must remain sterile. While packaging is not something ordinarily considered to be part of conceptual design work, I discuss the departmental interactions here because the problems that arose were constantly cited by Research and Development personnel as problems when newly developed products were produced in-house. Drawings and prototypes played a significant role in both this conflict and its resolution.

As the project moved from the early prototype phase to the pre–production prototype phase, federally mandated technical tests for bioburdens and sterilization qualities were performed on the new prototypes arriving from Fabrimed. The new prototypes also had to pass the same technical performance tests that the early prototypes had undergone. The product engineer points out the potential for communication gaps and technical slippage as this transition takes place:

We've already done a lot of tests in house, and we will have to take the initial prototypes coming off of assembly and do further tests and make sure that what we found with our initial prototypes, we're finding with our production prototypes. And that's what we need to find out. We need to put together some ideas that we don't have in this instrument [a "quick and dirty" prototype]. Because this was lathe cut, and now we're going to go into molding. So bringing over *one* concept, and you're switching it over to the next. So there are some experience things that need to be hopefully communicated from everyone when you switch. Any time you switch from one method of production to another, those things need to be communicated.

The first step toward moving the production of the Optimed Inserter from the auspices of the Research and Development Department over to the control of the Manufacturing Department, the usual and ultimate progression, was that a manufacturing representative began to attend the weekly design team meetings. Susan, a manufacturing engineer, began to attend design meetings shortly after the outside vendor had been chosen. Occasionally, her supervisor, Ralph, would also attend. Her participation consisted mostly of input regarding bureaucratic routines in ordering materials, labeling processes, and supervising inventory control numbers for the packaging of the new tool. She also was knowledgeable about medical clean-room packaging procedures and often asked for information and clarification about packaging quantities and requirements. The company scheduler and the Manufacturing Department arranged for extra workers to be hired to seal the instruments in double pouches, send them off for sterilization, label them with specific lot-coded labels, and put them in boxes especially designed for the product and its promotion. The manufacturing representative was also in charge of a team of workers engaged in developing a polishing technique for the stainless-steel tubes to be installed in the Optimed Inserter. The technique had to polish the tubes in a manner that would render them smooth inside, outside, and along the beveled entrance to the tube against which the Mylar paddle retracted as it was drawn inside the tube. Any rough edge or burr might tear the paddle or scratch or mar the expensive lens the paddle cradled during its retraction or release.

The tube-polishing project was technically problematic, and an outside vendor simultaneously worked on the same problem. Pressure began to mount as the Marketing Department promised customers that the new tool would be delivered with new lens orders. However, a polishing technique that met both quality and quantity requirements had not yet been developed. Thus, Fabrimed's manufacture of production pro-totypes—which needed to pass both federally mandated tests and physician-supervised empirical tests before the new product could be marketed—was held up by the shortage of sufficiently polished tubes. At this point, both the Research and Development Department and the Manufacturing Department were doing their best to work together to solve the polishing problem, while the tube vendor appeared to be making empty promises. Ultimately, input from the experimentation in techniques carried out at MediVis and an innovative approach by the vendor allowed simultaneous polishing of multiple tubes. Production output at Fabrimed then increased sufficiently for the extensive testing phase.

However, once the technical problem was solved, another problem arose. The Manufacturing Department, which had been very cooperative up to this point now insisted it had no personnel, no equipment, and no space to handle the packaging of the Optimed Inserter. It maintained that the additional workers who had been hired for the upcoming project had been assigned to another job during the delays and were now unavailable. They also balked at authorizing the transition of the official drawings for the project from Research and Development over to Manufacturing. Manufacturing representatives insisted that the drawings submitted by Fabrimed were fraught with errors and were missing important information. Essentially, the Manufacturing Department was using bureaucratic procedural policy—the prescribed trail of paperwork and its routing—as a tool to block the next step in production, even though they were not involved in physical production since the product was being fabricated outside the Manufacturing Department and outside the company.

Why had a cooperative department suddenly become a stumbling block? To explain the phenomenon, it is necessary to add some more data to the narrative to see how a prototype that was a boundary object, perceived from multiple views and through multiple interpretations, for the Manufacturing Department changed from being a positive conscription device that recruited management, employees, and departments to join the bandwagon into a potentially hazardous and costly enemy that had to be resisted for the good of individual careers, the department, and the company.

During the anxiety-ridden period of polishing problems, the product engineer took a quantity of the new prototypes to another ophthalmological exhibition and gave MediVis sales representatives their first opportunity to get their hands on the new, highly touted, and long-awaited tool. The first day was problematic as the overzealous sales force started to demonstrate the Optimed Inserter manufacture prototypes without benefit of instruction and managed to break several of the Mylar paddles by using incorrect loading techniques. On the second day, after the sales force had been trained and knew the correct way to load and handle the instrument, the Optimed Inserter's reception was highly successful. Physicians and salespersons were impressed by the new tool's performance and enthusiastic about its medical application. However, the reports of the first day reached the home office, and rumors of the instrument's embarrassing failure quickly spread throughout the company, especially in the Manufacturing Department.

The project engineer returned with the good news of the second and subsequent days of success, only to be greeted with reports of the embarrassing first day. But the harm had already been done: rapport with the Manufacturing Department had been sabotaged. It reinterpreted the instrument from a future success that they wished to participate in to a sure failure and headache for Manufacturing rather than for the designers who created the problem. Manufacturing management's concerns and interests colored their reading of the issues relevant to the Optimed Inserter's production.[5] In particular, manufacturing representatives consistently brought up a previous project in which design problems had been discovered after the product had already gone into production. Because of this late discovery, the Manufacturing Department had a backlog of orders it could not fill because its inventory was a faulty product. It was left with the double penalty of unfilled back orders and worthless inventory for which manufacturing management took the heat when they felt that the Research and Development Department was responsible for the situation.

The Manufacturing Department jumped off the bandwagon and became sticklers for protocol, insisting that all bureaucratic procedures be followed to the letter as the paperwork for the Optimed Inserter began to be turned over to them. Moreover, manufacturing representatives claimed there were errors in the drawings. It is notable that the objections leveled at the Optimed Inserter focused first on the drawings. During one design meeting, a manager in the Manufacturing Department attacked the drawings Fabrimed had provided for manufacture of the molds, stating that the drawings were inaccurate and had much crucial information missing. The political nature of this accusation is heightened by the fact that Manufacturing would not be using the drawings because it was not involved in the production process but was only packaging the device. Its acceptance of the drawings was only a bureaucratic formality. At a subsequent meeting, manufacturing representatives insisted that the materials designated in the design documents were incompatible and could not be bound to one another. When I asked a manufacturing engineer on more than one occasion to show me the errors in the drawings, she delayed. Several weeks later, after the intervention of top management, she informed me that the drawings had been misread. It took high-ranking, high-powered intervention to coerce the cooperation of the Manufacturing Department. Yet even after having been forcefully informed by top management that Manufacturing was to stop stalling and was to provide the needed workspace and personnel for the completion of the project, manufacturing representatives made sniping comments regarding the inadequacy of the drawings and the following of bureaucratic procedures when they took charge of the final drawings at the final meeting of the two departments. Whether misread or not, the example once again shows how the drawings as boundary objects can be interpreted not only from different technical viewpoints but equally from different political viewpoints.

Conclusion

What is at stake here is not trivial. Engineering sketches and drawings are the building blocks of technological design and production. Moreover, because they are developed and used interactively, these visual representations act as the means for organizing the design to production process and serve as a social glue between individuals and between groups. The cascade of ideas on paper, layered and refined from initial tentative sketch to final product of accumulated knowledge, which Latour (1986) has designated a center of calculation, is necessarily also a center of power—the locus of control and negotiation. Looking at the everyday process of work and the visual practices of design with attention to sketches, drawings, and prototypes not only reveals a power struggle but also clarifies the interests and actions of the actors and the process of the final outcome.

While visual representations as boundary objects and conscription devices can facilitate multiple readings and hence collaboration and consensus, the later are not givens but are the outcomes of specific circumstances in a design and production setting. Conflicting claims reflecting various interests also center on the paper and object pair, illustrating not only the flexibility of those actants as grounding devices for multiple interpretations but their crucial nature as conscription devices connected to power and control. Control of access to drawings determines who may participate in the design process—as is illustrated in the exclusion of the Manufacturing Department from the initial phases of the new design. Moreover, at such a center of power, control over the form that knowledge will take, control over the use or nonuse of that knowledge, the control of workers and work, and access to material and labor resources are at stake. Since visual representations are located at the center of power, they are the locus of action, which may be negotiation and consensus or it may be conflict and power plays.

In this case study, the visual representations are engineering drawings and prototypes. However, these are not the only forms of visual representations that help organize time, thought, work, people, and politics. Examination of the construction and use of visual representations employed in many other kinds of work can reveal how access to information is controlled in ways that discriminate against some and empower others, how group cognitive work and its outcome is organized, and where centers of calculation and hence power are located in an organization or community. The findings presented here suggest the fruitful potential of attention to the role of visual representations in areas beyond science and technology studies.

In the next chapter we examine what happens when the mode of generation of documents—so vital to work, workers, and production relationships—is significantly altered, as we look at the many interlocking restructurings brought about by the implementation of computer graphics.

6

Computer Graphics in Design Engineering: How and Why Changes in Visual Culture Cause Restructuring

Chapter 5 showed the centrality of drawings to both consensus and conflict. When changes occur in the way such visual work is done, they necessarily spill over into the relationships around that work. Implementing a graphics system tied to an overextended database, for example, destroyed informal ways of doing work and led to a shop that lacked the literal nuts and bolts needed to assemble a new turbine engine package prototype. A second significant change at this site was the insertion of the CAD drafting department between design and production, which restructured the paper flow between departments so that errors detected on the shop floor were referred to the CAD drafting section for correction rather than to the designers, setting up the potential for further error. This chapter examines the kinds of restructuring that computer graphics have engendered and looks more broadly at such proliferations of restructurings.[1]

American industry has put a great deal of faith in computers and in computer graphics, in particular, to deliver the promise of automated design-to-production robotics that will restore U.S. industry to a world-leadership position or at least a position equal to its Japanese competition (Downey 1992a). However, the idealized view of computerized graphics is that one best method can be applied to all situations—a problematic premise, at best. One method, one machine, one way of doing the job simply will not fit the actual practices involved in the daily activities of design work. As several researchers have documented, design engineers do not accomplish their work using design-school protocols, which are formalized routines akin to the notions incorporated into graphics programming (Ullman, Stauffer, and Dietterich 1987; Bly 1988; Tang and Leifer 1988; Sinclair, Sieminiuch, and John 1988; Vincenti 1990; Henderson 1991a, 1995a, 1995b; Law 1992; Bucciarelli 1994). Despite the thoughtful work that went into the design of computerized-graphics software and hardware, the attempt to make this one technology fit different kinds of industries, different kinds of engineering, and different kinds of companies, each and all of which have their own ways of getting the job done, has resulted in all sorts of restructuring—from the way an individual puts down a line to relations between workgroups and companies. Such restructurings tend to be exponential: they proliferate more and more restructuring and expand to almost every aspect of generating design and putting it into production.

Of these multiple and interlocking restructurings, first and most dramatic is the restructuring of the jobs of designers, drafters, and engineers. Second and most recognized by those using the technology is the restructuring of the design process itself that results when CAD systems are implemented where they do not quite fit. Third, and perhaps most intractable, is the potential rethinking of drawing standards on multiple levels. The different standards needed in various design applications conflict with the assumption that standards of CAD machines can fit all jobs, raising the question of how to build in flexible standards or how to create new standards to analyze CAD work. The problem is complicated by user disagreement concerning optimal standard flexibility since individual, company, and industry interests are involved.

A multitude of unanticipated problems have occurred with the introduction of CAD that designers, drafters, and engineers have had to find ways to accommodate. Among these problems are the loss of the informal practices of the paper world which facilitated in getting the job done in an expedient manner, as seen in the turbine package study, and the difficulties of maintaining faith in an output in which the traces of process are missing. All these restructuring issues are related to the central one—the cognitive and practical changes that occur when designers move from physically putting a pencil on paper to instructing a computer to generate a line. The focus here is on how such restructurings impact engineering design in an aggregate way.

Previous chapters have documented how sketching and drawing practices serve as the basic building blocks of technological innovation and shape how the work is structured, who may participate in the work, and form, format, and content of the final products of design engineering. In response to the massive restructuring brought about by the implementations of computer graphics, designers and drafters create mixed-use practices that incorporate the new technology with older paper-world strategies that include sketching and drawing. These mixed practices, which are discussed in detail in the following chapter, are not merely emblematic of change in process but illustrate the importance of the tie between skills and action-based knowledge. The resilient and innovative mixed-use strategies of design engineers, drafters, and their managers get the job done in a work environment fraught with changes that can denigrate their work and bring about a loss of pride and status in their jobs as machines are given more and more credit for design work. The concepts of boundary objects and conscription devices, both of which facilitate shared meaning and simultaneous multiple readings of a visual document, underlie the analysis here as an aid to understanding the dynamics of group visual communication, creativity, and problem solving in design work. They also help reveal the resilience of practices embedded in a visual culture that serves as the source of mixed practices able to take advantage of new electronic-world tools while retaining the assets of paper-world ones. While mixed-use practices cannot solve all the restructuring problems generated by the implementation of the new technology, they come to the rescue in accomplishing work goals and, perhaps more important, they point up that more is going on in the work process of design than the linear approach of computerized algorithms.

Methods

Because it would have been possible to form a distorted view of CAD systems and engineering design from only the two case studies presented in chapters 4 and 5, I conducted in-depth, open-ended interviews with engineers, designers, drafters, and managers from approximately twenty-five other companies to better understand the universe from which I was sampling. The majority of these industrial sites were located on the California coast from just north of the Mexican border to Silicon Valley, though one site was on the U.S. southeast coast. The choice of companies was based purely on where I could obtain access, since most businesses are adverse to allowing visitors anywhere near new design work. In many cases, I conducted interviews in lobbies, lunchrooms, outdoors, or homes. Even when I could conduct work-site interviews, corporate security would often not allow me to use a tape recorder. Besides pestering friends, faculty, and colleagues for contacts, I used the alumni association of the University of California to locate graduates working in engineering firms and then used a snowball technique to find individuals involved in computer graphics and design. In one case, the link was a group of chip design engineers in Silicon Valley who played softball together. These techniques were adequate for talking to design engineers but less successful for finding a large variety of firms using computer graphics. Through the graciousness of a computer-graphics consulting firm, I attended a workshop for graphics systems managers where I made more contacts for worksite interviews.

Since this chapter, unlike the case studies, is based on interview data, it has a different texture than the chapters based on materials gathered through participant observation. This means that both researcher and reader must trust the word of the informants without data based on the researcher's observation of daily practices and events. Not only is a different methodology used here, but the issue of trust, or the lack of it, came up frequently in the comments of graphics systems managers cited in this chapter. These managers found that experienced drafters and designers felt disinclined to trust calculations that they themselves had not made during the process of rendering a drawing. While we must trust that the observations and judgments of the informants, who are first-hand experts on what they do every day, were accurate and that the researcher asked the right questions for the validity of the information presented here, designers and drafters must trust the calculations of the machine that has become the new expert, rather than their own personal, action-based skills and practices.

It is important to note that, at the same time, those who have developed their skills using the new computerized drafting tools can find their practice-grounded ways of knowing at odds with standardized conventions grounded in the paper world in which drafting developed. The example given in chapter 3, taken from the complaints of design instructors in the engineering school at a major university, shows how a new visual culture is beginning to emerge through the practices of engineering design students trained on CAD systems. I review it here to underline the connection between actual work practice and ways of seeing and thinking. Engineering design instructors complained that many of their students trained on CAD systems consistently made an identical error in electronically rendering a cylinder. Since CAD-trained students did not use the construction lines that paper-trained designers use, no line was present to suggest the crucial line across the cylinder's opening, a convention that designates the figure as cylindrical rather than flat. The line at the cylinder opening had become an arcane detail instead of part of the drawing practice for computer-graphics users because the figure could be constructed without the use of construction lines by simply employing the computerized mirror function. While the absence of the crucial line is still marked as an error on the students' work, the ubiquity of its occurrence points to the emergence of a new visual culture tied to computer-graphics practice for a generation of designers trained on graphics software. This newer practice-based visual culture will influence the way electronic graphics users conceptualize visually, and it will vary from the visual culture of the paper world. This is just one example of the kinds of changes occurring when individuals use the new technology. Other restructurings reveal other issues and present other problems. Beneath all such outcomes is the problematic underlying assumption that one method can fit all applications.

The Problem of One Best Method

CAD systems are designed for limited methods of design work, and the methods chosen for CAD work are likely to fit the needs of the company they enter only imperfectly. This does not mean that the designers of the systems do not think about users. They just have limited models of users and the problems of use. Assumptions that there is one best way of doing design work are mistaken, and views that hold that such a method can be captured in a computer-assisted-design system that will fit all needs are overidealized. To illustrate these points, here are two stories reported from two different sites where the same software was being employed with very different results. At one site that I will call Ramtech, a large and well-known computer firm that produces hardware and software for industrial use, a very profitable interaction developed between the designers of the computer-graphics program and the hardware designers who were using it. One of the hardware design managers described the complementary interaction:

Since we're both Ramtech, they're [graphics-design software programmers] real interested in our feedback. So the division and group managers, high-level managers from that group, have come over and had staff meetings actually at our site, specifically to sit down with users and ask them what they think. And we get a lot of feedback directly into their system. And they're changing a lot of their programs because of that. They go over and they say, "Hey, this doesn't solve their problem. Let's fix it." Well, there's a lot of interaction in that regard—our division usage and how we feel about it to what they work on it.

Asked what kind of problems both teams approached, he replied,

Primarily interface type things where you want to be able to go from a system A to system B to system C and not lose information in that process, and we want that to be a three-command sequence kind of thing rather than 400 commands.

Okay? In other words, we want it to be very, very simple in how that interaction occurs.

Having covered the necessity of simplicity of commands, he added some comments on the insufficiency of parts libraries:

They didn't have sufficient part libraries. What the electrical engineer would like [is] to say, "I want to put an A to D converter here," boom, and have an A to D converter show up. Only he doesn't want just an A to D converter. He wants to be able to pick out part numbers that are exactly, you know, what he would be buying and have those be there so it would show that he has four A to D converters, maybe in one package for A or whatever it may be. And those libraries were just real limited.

A systems manager at the same site pointed out the interactive and ethnographic character of the graphics program development research:

So that was the intent, to mimic what's happening in real life, and how do you do that? Well, you go to real life, and you watch, and you observe, and they've done a very good job of that. See, the tendency is actually to think about what would you like to accomplish and then what's the easiest way to accomplish that as far as your software is concerned, which may not be the easiest way to accomplish that as far as a person [using it] is concerned. And that's very, very common in software, and that's what—that's when you end up getting away from user friendliness. But they wanted to take the opposite approach. They wanted to make it as easy as possible for the person. . . . They talked to [designers] about how do you go about doing this? If you wanted to form an image, how—what's the easiest way for you to do that, and so forth. . . .

For example, when I want to draw a part, there's things known as construction lines which basically you draw at different dimensions. And then once you have all your construction lines down, then you actually like trace along certain construction lines, and you end up with the shape of your part. And then you'd like all the construction lines to just go away. That's how people typically do drawing. And they took that same approach for doing—for drawing these arbitrary two-dimensional figures. So there is kind of a standard technique that really isn't that far from how everybody does it. I don't think anybody—. I think doing it this way doesn't really crimp anybody's style, and if it does, it's probably better that they learn how to do it properly [laughs]. . . . Generally, when you're on the CAD system, you don't worry about perspectives. What you do is you do whatever's easiest, and then you let the system generate a different perspective, if that's what you want.

The hardware designers at Ramtech are saying essentially that the software designers acted like ethnographers and came to study how designers really worked and then modeled the computer-graphics system to fit their needs. But that does not mean that the same software can necessarily fit the needs of all designers everywhere. Returning to Selco, the site of the first case study, the following comments of the packaging section design manager discuss his interaction with Ramtech and their graphics system:

On two occasions I've bought programs. They've come right at the end of the year. They've got somebody that says, "Look, they've got a special. You can buy it for this price. It's normally forty grand. It's wonderful." They give you a quick flash demo, show you the bloomin' space shuttle doing its wobbles and everything. You say, "Ah, fantastic. You know, this is just what I need," and get it for twenty-six thousand dollars. I go and ask our computer experts, "Will that run on the VAX?" "Oh yeah, yeah, yeah. No problem." Then you get it into the VAX, and you load the whole problem up. Then, "Well now. How do you want to put the data in?" "Well, what do you mean? I have this vibration monitor. I want to collect this data and put it in?" "WE—no, no. You can't do that." Then you gotta have another data software acquisition system. They nickel and dime you. I am now so skeptical I don't buy software packages any more. They will sell them to me on a turn key, and they will come in and make them work. That's why I don't do much business with Ramtech, 'cause Ramtech won't turn key their jobs. They'll sell you the software and the hardware and leave it up to you to make it work. And then they'll quote you some outrageous price to sell you an integrated system and that's the problem. I mean, I know Ramtech makes beautiful instruments. They do. But their ability to sell you an integrated package is the pits.

The complaint here is not just that Ramtech will not "turn key" the job—that is, come in and make sure their products perform at the individual company site—but, more broadly, that systems like this in general do not easily fit every company's needs. This manager went on to point out that this is not an isolated problem with just Ramtech. Nor is the problem one that occurs only in industry but is actually the same problem of incompatible systems and software (and the escalating costs of attempting to integrate them) that many individuals face with their own personal computers. The ethnographically designed perfect-fit CAD system fits perfectly only when it is developed within a company for its particular needs—and this kind of fit is the exception, not the rule.

Returning to the earlier example of the emerging electronic visual culture of CAD-trained engineering design students, it is interesting to note that the Ramtech system manager's assumption that the "standard technique" can withstand changes in practice conflicts with the CAD-trained students' tendency to disregard what they regard as inessential construction lines. This is because they can more easily use a computer function to attain what they perceive as the same end. While the system manager is no doubt correct that "it's probably better that they learn how to do it properly"—that is, draw in a manner consistent with paper-world drafting conventions—he does not anticipate how changed practices will challenge those conventions.

At sites that do not custom design software to fit their own specifications and that have people use graphics software packages that are not their company's product, people still get work done with computer graphics, but only while struggling with many obstacles. They must manage the set of interlocking restructurings that the implementation of new graphics systems engender in the workplace. One of the most basic is the change in definition of drafters' job specifications.

Job Designation: Computer Operators versus Drafters

Knowledge and Ownership: Changing Domains of Authority

One of the most dramatic forms of restructuring that CAD has brought to the workplace has been the reorganization of job designations around the design process. The most significant of these changes has been the diminution or redefinition of the role of the drafter. Early CAD-CAM promoters promised to deliver automated production in a direct line from the computerized drawing, generated by the engineer, to the automated machining and assembly of parts. This model intentionally left out the intermediary work of the drafter and thus eliminated the intermediary step of finalizing polished designs. Companies that attempted to do away with their drafters almost immediately ran into problems because most engineers do not want to be bothered with the painstaking detail needed for a finished design and computer operators lack the tacit knowledge and skill of experienced drafters (Saltzman 1988, Majchrzak and Saltzman 1988, Manske and Wolf 1988). Informants at numerous sites indicated that skilled drafters are needed to train and oversee computer operators in order to maintain clear communication between engineers and the shop.

Drafters' work authority has diminished almost everywhere. They may now be seen as the equivalent of computer operators, whose successes are credited to the machine. Their reliance on computers may, in fact, be real enough. Many mathematical calculations are now done by the machine rather than on paper by drafters. Other parts of their traditional expertise, notably their knowledge of materials (how far something can be bent without weakness, for example) is now redefined as being in the domain of the engineer. A consequence of this gutting of the drafter's role is that in a good number of companies, especially in Silicon Valley, drafters are now cut off from advancement to engineering. With few exceptions, no longer can someone who came into the company as a drafter work up the ranks in design departments to an engineering position. In a few companies, the original vision of CAD-CAM is still embraced, and engineers are paid professional salaries to do painstaking detail work. In other companies, computer operators simply digitize the work of engineers into the graphics system, losing the expertise of the drafters. In still other companies, the situation is some sort of mix between old and new responsibilities. Experienced drafters and designers tell computer operators what standards to use and what materials specifications are needed so that workers in the shop will understand the drawings and can get the job done. The latter arrangement appears to be the most efficient, yet despite this, it is rare. The crucial nature of drafting skills tends not to be recognized, and rewards and acknowledgment for drafters have dissipated. The result is that drafters have less pride in their work. They feel less control and connection to the work since the traces of their contribution in the process are now being lost.

Changing definitions of the drafter necessarily affect the training of new drafting personnel. The drafting supervisor at a leading computer company that is called Orion here points out the changes in the knowledge attributed to computer operators, drafters, and engineers:

What is a drafter? I've questioned that myself. I've had a lot of companies call me up and question, "Do you have the same idea that we're having that we're no longer getting drafters, we're getting computer operators, out of school?" We find that a lot of the knowledge that goes along with drafting, understanding the thread size of screws, understanding plating materials, understanding materials themselves, um, understanding relationships between the two materials—that has gone away. That is now an engineering level. That is now an old-seasoned-designer level type of knowledge. And that in itself has an impact on productivity, has an impact on structures of CAD groups.

His statements that former drafting expert knowledge is "now an engineering level" or "an old-seasoned-designer level" knowledge illustrate the changing domains of authority. He stated further that many of the CAD drafters that Orion now uses transferred into the Drafting Department from all sorts of other departments in the company and then learned how to operate CAD workstations. They had little or no drafting experience. He complains that while these people know how to operate the system and make a pretty picture, they do not understand the meaning of its component parts:

They copy a file from a note—standard library of notes—they copy that. They know, "Well yeah, I have this plastic. These are the standard notes we use for plastics here at Orion. It tells everything.'' It tells you about the materials, the draft angles; it tells you about cosmetic specs, the textures, and everything. You know, the whole gamut of what's required, the whole process. Well, it's an automatic thing for the drafters to take that. Have that drafter read that. Have that drafter understand those notes is a whole new ball game. Three-quarters of them, if not more than three-quarters of them, don't understand . . . even some of the old timers. Because everything is there for them and it's no more of a decision on their part. It's a matter of selecting something. That's a problem we have, and it's not only here at Orion. It's in other companies too.

He added that such drafters today do not really understand their job because they do not participate in the process of building up the drawing:

I don't think I would get one drafter that could do all the notes and do it correctly because he's never had to deal with the process. He's never had to deal with understanding how this thing goes together or what's required for somebody outside to make this part.

In other words, he is saying that the knowledge of the new computer operators is not grounded in practice.

Besides the loss of materials knowledge and experientially based knowledge, the drafting supervisor also lamented the inability of the new CAD drafters to do the calculations that justify their work because the computer now does that for them. He made the point, once again, that someone who has the experiential knowledge is still the middle link:

So CAD has become a very effective tool, and it has become a real concern of a lot of professions, I think, because it could have a real adverse effect on what you get done or the structure that the company has to have in order to complete things. I think it needs to be addressed on a level that now we have a new wave of people that are now CAD operators, but are they drafters? Now, do we have an intermediate level of liaison people from engineering to drafting, people who define? I find that I am that person here at Orion. I define a lot of the engineering input to the drafters. I define a lot of the engineering input to the manufacturing world. I do a lot of the intermediate tooling and design or support that has to be done because it goes away. There's a lot of it missing there. A lot of the drafters rely on me to do a lot of their decision making or to help them out in a lot of things. . . . Everything that's done has to have some questions answered.

The remaining question is, What happens when those who have the knowledge retire and the senior CAD operator, who only learned which notes to use but never why, comes to fill this intermediary position? The Orion drafting supervisor thinks a further line of stratification will develop between CAD operators and CAD drafters:

Are you a CAD operator, or are you a CAD drafter? I think there will be a definition between those two. I think CAD operators will be duplicators and tracers like we used to have on the board. I think CAD drafters are going to be seasoned drafters that now learned how to do CAD things on CAD, and that hasn't happened. The technical part of the job has to be taught more, more so than teaching them how to use the system to put lines in place. Drafting needs to be taught. . . . They could be drawing things that are bent and going to break in the sheet metal shop. . . . You deal with the mean of the material in order to understand the bend factors and whether you're going to have stress points or whether it's going to bend a certain part. . . . Materials have to be determined. All that has gone away. Those were concrete things, foundations to understanding what your job was all about. Things have gone away.

But beyond the development of new stratifications in the CAD drafting world, he also notes the personal loss of self-image for experienced drafters:

I think a lot of the pride in the job has gone away. . . . I go to the American Institute of Design and drafters—they have conventions—you know, a membership. And if you go to a convention, the old timers, you know, they'll sit there, these guys, 'cause they've been around, with white beards. They still put their coffee cups on drawings. Those are the kind of guys you really love—you know you really hated. And, ah, they just say, "Don't even talk CAD drafting to me. There's no such thing as a CAD drafter. Either you're a drafter or you're a CAD operator, but there are no CAD drafters." And they say, 'cause every CAD drafter they've ever seen "don't know what drafting is." . . . I think it's an art, a profession that's gone away because of technology.

This drafting manager is no Luddite. He simultaneously acknowledges the efficiency that computer drafting contributes and yet realistically stresses that checkpoints and boundary work are needed to make the new system work so that information crosses over from engineering to production in an understandable manner:

It's [CAD] great for efficiency at certain levels. Like I say, it's got real benefits. It short cuts a lot of the process. By the same token it puts—it requires a lot of the intermediate levels, requires a lot of checkpoints. . . . It takes us creating a lot of intermediate structures to support a lot of things, and that's not just here at Orion. It's in a lot of companies.

Unlike the policy of the old-line, turbine-building, heavy industry of the first case study, where drafters can still work up to designer and then to engineer, throughout Silicon Valley drafters and CAD operators cannot cross the line into engineering without a university degree. This new philosophy had impacted the drafting supervisor himself, and he pointed out the inconsistencies of such a policy when drafters are sent in to do part of the job of the engineer but are denied the opportunity for advancement:

Drafters cannot, can no longer go to the design world. There's really a locked door. We have drafters working there. This is kind of two-sided here. They'll call me up, and they'll ask me to send drafters over to sit by the designers or to sit by the engineers to actually—to actually go in to work on their CAD system . . . and ask them to do part-detail. But that individual is not qualified to work full-time in that organization because they're non-degreed. Orion has taken a different view of their engineering disciplines. . . . I think it's very detrimental to real-time stuff being done over there. Some of the things we get from product design, it's not—it's not good to work with. Some of the files we get, the CAD files, have to be actually redone. We can't even pick it apart and use it in any kind of a way because it's so—so badly generated, so badly—no, no standards or anything of that nature. A lot of the fine detail is not even there because that person has so much knowledge of what they're conveying that they forget that the person they're trying to convey the message to doesn't understand what they understand. . . .

   Myself, I don't qualify to be a product designer any more, although I have twenty-two years in the design and drafting world. I've designed main-frame computers . . . and I've built them right next to my design board that I design them on, yet I don't qualify because I don't have a degree. I have years of technical schooling, but I don't have a degree, so its kind of unbelievable. Yet they'll call me if they have problems and concerns about processes, how to do things, what should be done, certain things. They know exactly who to call.

Just as the job designation for drafter versus computer operator has become ambiguous, so too has the actual work itself been restructured.

Restructuring Design Work for the Individual

Loss of Informal Short Cuts

The restructuring engendered by graphics systems is most palpably felt at the micro level of interaction between individual and machine as the design is constructed. This restructuring of design work can be traced both in the process—the activity of design work—and in the product—the visual representation itself. The whole justification of the time trade-off for the completion of the computerized model is supposed to come in the accuracy of the model, which should cut down the high costs of correcting errors in the production phase. Fudging—not following the dictates of the machine and resorting to paper-world short cuts—compromises this capability. One CAD manager mentioned the injunction against fudging so that CAD data will transfer across systems without distortion. It is important to note that fudging does not imply poor workmanship on the part of designers and drafters but rather has been one of the informal conventions of the drafting world. Such conventions were quite adequate in cases where the actual mathematical calculations necessary for a peculiarly curved line would be tedious to perform but not of particular use since the machine shop would use the consistency of the grinding machine and merely follow along the drafted, curved line to produce the requisite shape for a mold or piece of machinery. The drafter would use a template, sometimes a malleable one, to trace the curve without calculating every point along it, a useful and time-saving routine. However, such manual tricks of the trade are not allowed when using CAD, which requires that each and every mathematical point along a curved line be calculated.

Another problem is that the wire-frame format on the computer screen can be so confusing that those skilled in the practices of the paper world often prefer to use the manual override to render things in a format familiar and comfortable to them. The dictate that no fudging take place is a double penalty for those experienced in the paper world. First, they must abandon accepted short-cut practices that have aided them in getting the work done in the past, and, second, when the work becomes confusing, due to the overlapping lines of the wire-frame format, they cannot turn back to the well-known cognitive territory of the paper world, in which their visual knowledge can help them literally see their way through a problematic area. Such visual skills are less valued when trust is put into the machine instead of the person and when action-centered knowledge born of putting the pencil to the paper becomes subservient to skill in operating the computer.

Computer Skills

To use CAD properly all the way through means developing new skills particular to the new technology. Zuboff (1988) points out that a compensation for the loss of action-centered skill, such as drawing on paper, is that those who now use computers instead of hands-on involvement develop what she refers to as *intellective skill*. According to Zuboff, intellective skill is a new set of competencies that combine abstraction, explicit inference, and procedural reasoning (1988, 75). While this is true to an extent in the world of computer graphics, the new skills are so essentially different from the old visual ones that they bear almost no relationship to visual knowledge. When manual drawing skills are replaced with computer-use skills, the ability to think visually needs to remain paramount if designers and drafters are to maintain a grounding in their work. A Metrotech piping design supervisor noted that his designers are so involved with the new skills required to do their job on the computer that they appear to have lost sight of the total picture. They no longer come to him with suggestions for the total layout. He pointed out that his civil engineering drafters and designers used to make suggestions, such as advice about moving or rearranging the piping for a large and complicated installation to shorten distances and avoid other structures in the layout. He feels that because of their intense interaction with the computer, these designers have lost sight of the total picture. Designers may develop new skills oriented to computer-graphics use, but they may be gained at the cost of more visually oriented skills. When I asked designers and drafters about new skills they had developed using computer-graphics systems, they gave answers about computer techniques—developing a little program or a macro to do something—but nothing having to do with visual elements of design.

A problem of a different sort involves computer skills with macros. The Orion CAD supervisor pointed out that the graphics system his people were using at that time did not automatically update all the related files when a change was made in one file. However, one of the non-degreed designers used his computer skills to write a macro that would accomplish the updates without having to go back in and change each file. As important a tool as the macro could be for everyone in the department, the individual who developed it did not share it willingly with his group but gave it to individuals only if they asked him for it. The supervisor pointed out that even though management tries to instill a "please share, we will recognize your accomplishments" philosophy in the department, given the loss of recognition for drawing skills, people are protective of the new skills they develop:

Everybody does their own little part of the design and does it their own way best. They may not want to share: "If I'm good, I want impress the boss and get good work. I'm not going to let this guy know what I'm doing so I have to share the good work." We try to get designers to share and we share with them. We showed them how easy it is to use CAD. Sometimes they listen. Sometimes they ask, "What do you guys know?" There is a lot of ownership in the CAD world. We have a good network. Everyone, if asked how to do this, two or three will help. We asked him for it. He said he would write it down if had time, but he didn't have time to show people. The manager said, "Okay, we'll figure it out. We'd like to have it. If we work on your files, someone may have to change them and not know what you did." . . . That sold him. So we manipulate ambitions around here.

As justification of jobs becomes an issue and recognizable individual skill in drawing becomes less valued, those using the new technology try to protect the new skills they learn. Further restructuring-generated problems occur at the level of the drawing itself.

Restructuring Design Work Itself

Restructuring relations of those involved in design work and restructuring the processes employed in design work can both be traced in the actual drawings. In order to discuss the drawings themselves, it is helpful to turn to the language used to critically discuss other visual representations—the language of art criticism, which encompasses elements such as aesthetics, style, and signature. At the same time, it is also necessary to retain the language of engineering, in which standards and standardization play a prominent role. Aesthetics and standards are equally important in engineering, and both can be read in functional and symbolic terms. The line between aesthetics and standards is drawn mainly by the distinction that standards embody the specified rules of design work while aesthetics encompass the unwritten rules that distinguish mediocre work from quality design.

Aesthetics

In the world of visual art, aesthetics have to do with critiquing and judging visual phenomena and in setting criteria for making such judgments. Originally this was done in terms of that slippery word *beauty,* but today more diverse categories are used. While engineering drawings can be judged beautiful, even by art-world standards, aesthetics in engineering design are not merely issues of beauty but rather are the internal criteria of quality judgments. Two aspects of aesthetics that were mentioned on several occasions by informants in this study were aesthetics as functional and aesthetics as symbolic, though these were not mutually exclusive. Both were recognized as elements indicative of individual and company quality and style.

   Several informants mentioned the functional aspects of aesthetics in terms of neatness and organization. A civil engineering CAD manager pointed out that the implementation of computer graphics has resulted in two extremes that he classified in terms of aesthetics—sloppy unfinished work with no attention to detail versus too-refined and hence too-expensive work for a given situation:

I have seen two tendencies, um, a few more—the tendency to believe what comes off the computer is true so there's a little less checking the fit of the parts. Another tendency is to be a little sloppier on CAD than [people] have been on the boards—not to do a drawing in a very clear or very consistent manner. I've seen people issue drawings coming out of CAD that they would never have issued had they sat at the boards and drawn it by hand themselves.

He added the following example:

Now, in a manual mode we are very careful to choose the appropriate scale or to choose the way we plan our notes. Because if we don't—if the project engineer doesn't like the clarity, he'll tell the guy to erase it and change it, which is a problem. It's a hassle, and people don't like to erase what they've done and redo it. Whereas on CAD things are so easily changed that they tend to be sloppier about selecting a scale and selecting an appropriate way to show notes and so [are] sloppy or lax in our issuing those drawings. And so drawing a lot on CAD [may result in work] that has notes all cluttered together on top of what's being built. And a person who's reviewing [it] will say, "I can't understand what's going on here at all," whereas that sort of thing would never have gone out on the boards.

He pointed out that even though cleaning up the work is fairly simple to accomplish, such cleanup is not always done for each review session because of deadlines. He then described the other extreme—the overworked drawing that becomes too expensive for its intended purpose:

I've also seen change just because change could be made. It's like when we all got word processors, you know. We all made that final revision even though if we had typed it on the Selectric we would have said, "Ah, it's all right." So the same thing happens on CAD. Just because you know that change is fairly easy to make and you can spit that plot out, why, we try to make changes up to the last minute. So I've seen that too.

Not just neatness but orderliness in organization was another criterion often mentioned, especially in electronic engineering. A chip designer at Orion Computers described how orderly spatial arrangement of components within the chip should be constructed in patterns that are not only pleasant to the eye but communicative of their functional relations:

Well, there are no—as far as I know, we don't have any set of specifications for how our drawings should look, at least for schematic in the chip, um, but there are some kind of unwritten rules on how neat it should look and organized and how you want the—the way the schematic flows. The way things are placed on it should try to somehow represent what the chip does. . . . You should try to make the thing organized. And I've seen people just put stuff completely randomly, um, without any care for anyone else who ever has to look at it, later, and figure out what it is. If you put—. You could take all the parts of a car and place each one of them on a piece of paper and put a name on each one, a number, so you know which one connects to the next one. You could do that, and you wouldn't know what the heck you've got. You just see all these pictures everywhere. And some people do that. The other thing you could do is you could kind of draw the car and show how everything works together. That's what we try to do. We don't have any specific standards, there aren't any ANSI [American National Standards Institute] standards or mil. spec. [military specifications] standards. There aren't any of that kind of stuff.

The concept of flow, mentioned above, was also noted as significant by several electronic design engineers. At Astrotech a design manager pointed out that all the inputs should be on one side and the outputs on the other side so that "it looks like it flows." He added that electronically this convention makes no difference at all but that it makes drawings easier to read. Another chip designer at Techlink said that flow is a U.S. Department of Defense requirement and that when engineers do not follow the practice they end up having to change their chip designs. He added, however, that it is not officially checked for and that it is up to the engineer to incorporate it. Questioned further he elaborated that an "ugly" chip design is one that "doesn't flow well" or does not exhibit a "logical flow" and that a good chip design is one in which the output on one page should be input onto the next page. His other examples of an "ugly design" include wires that are named but not drawn or a design that is broken up, without apparent plan, over several sheets of paper. His neatness and organization priorities are similar to those mentioned by the Orion chip designer.

Style and Signature
Another aspect of aesthetics is identifiable style or signature, at both the individual and company levels. The most obvious element of signature is at the company level—the company logo and company standards, which are discussed in the section below on standards. While the issue of signature in industry may seem trivial to the uninitiated, it is as important to those in design work as the identifiable style and signature of a recognized artist. Traditionally, design schools, architects, and companies or individuals involved in design work have used a particular style of lettering, all their own, which becomes their literal signature, identifiable throughout their work. A civil engineer explained what is gone and what is retained with CAD:

When we were doing drawings by hand, there was a very definite personal signature of the individual in a drawing. I could look at a drawing without looking at the title box, and I could tell you who did the drawing in a lot of cases, just because I know their style over the years and how they letter or the certain way they show things. A lot of that has gone away now. I can still look at a drawing and still know who did it, just because I know the way certain people show things or I know which people go to an extra effort to make things look better. But lettering style, now, for example, has gone away. That's a CAD font now, so everybody's looks the same. So that's one thing that's changed.

To illustrate his point on recognizable style beyond lettering, he turned to the whiteboard in his office and drew as he talked:

Let's say one guy will show a piece of pipe as several thick lines, (drawing) like so. The fittings are thin lines. Another guy might show that same product as two double lines with thin lines. This [first example] looks much better in the drawing because the contrast in the line weight will show up to the contractor as, "Oh yeah, that's where the pipe is." In a busy drawing where you've got lines running everywhere, it might not be apparent where this pipe is in comparison to something Structural is trying to show about floor length. And so if this guy has done a good job of throwing this pipe, . . . he's selected a nice broad and a nice-size pen, with this type of thing I can generally tell who has good style and who doesn't have good style. There's some people we have in Civtech where I consider their work to be almost artwork. And in that part it's a battle to get them not to do artwork because they're spending so much time on their drawings that there's more dollars in the drawing than there should be.

The same manager added that designers try to maintain such levels of distinction even when using computer graphics:

It's very definitely in a hand drawing, and I've seen the same trend in CAD. . . . See, since the personal signature is starting to go away, so people are saying, "I want my work to stand out." So in certain areas they're going way extra in CAD to make sure that when somebody looks at their drawing they'll know that it's their drawing.

Hence, computer-graphics implementation can result in overly detailed and precise drawings as designers attempt to compensate for the loss of personal stylistics, but it also can result in the abandonment of attention to detail and appropriate scale for intermediate deadline presentations because these factors can always be corrected later. Those who are aware of the subtleties can tell the work of one designer from another even when he or she is using computer graphics due to aesthetic choices such as whether the input and output go smoothly from one page to the next, described as flow in electronic engineering, or whether clarity is provided by line-weight choices in mechanical and civil engineering. These are just a few examples of the various criteria viewed in their respective disciplines as evidence of good style or as symbolic of quality design work.

Drawing Standards

In engineering practice the concept of standards has more than one meaning. On a company level it can refer to a firm's particular ways of rendering, which thus become recognizable in the same manner that signature does. An example would be the use of very articulated symbol sets that realistically resemble the item represented—for instance, a pump symbol that resembles a pump instead of an abstract symbol that merely stands for a pump and must be deciphered from a legend such as the converging triangles mentioned in chapters 3 and 4 (see figures 3.6 and 3.8). Such realistic drawing practice may be regarded as a mark of quality workmanship because of the concern for clarity. Company standards may also designate certain typefaces or fonts or line thickness for particular uses or simplify a larger set of standards down to a manageable pool. One reason computer operators who were hired in many firms failed in the attempt to replace drafters (Salzman 1988) was that they had neither drafting background nor knowledge of the particular industry or company conventions and standards. Most CAD managers acknowledged that while computer-assisted design systems help in keeping standards, system operators still need a drafting background. In the following quotation a manager makes the point that company standards help simplify the potential mind-boggling array of standards set by the American National Standard Institute:

We try to adhere to using bold object lines to differentiate between object and dimension lines. Ah, we minimize the amount of dash lines or phantom lines or invisible lines. We try to make the drawing as readable as possible, as complete, to convey the message. We now have created standards for certain things in our CAD so there's no change from one drafter of them looking one way and another drafter laying out the next way. . . . We make all the details the same so they're on file. So our vendor doesn't get a PC panelization board that has a lot tooling, a lot of routing and shared points in it, or what we call scores. . . . We try to maintain some consistency. . . . We standardize our notes.

ANSI standards are more universal than company standards. They are an international visual lexicon so all-encompassing that subsets can be used by any industry to explicitly articulate such things as parallels, perpendiculars, surfacing techniques, and how closely machining must match the dictates of the drawings (figure 6.1). The Orion drafting

AMERICAN NATIONAL STANDARD
DIMENSIONING AND TOLERANCING

ANSI Y14.5M-1982

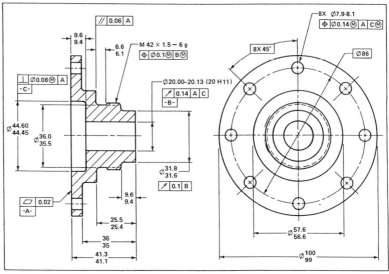

FIG. 85 FEATURE CONTROL FRAME PLACEMENT

AMERICAN NATIONAL STANDARD
DIMENSIONING AND TOLERANCING

ANSI Y14.5M-1982

	TYPE OF TOLERANCE	CHARACTERISTIC	SYMBOL	SEE:
FOR INDIVIDUAL FEATURES	FORM	STRAIGHTNESS	—	6.4.1
		FLATNESS	⏥	6.4.2
		CIRCULARITY (ROUNDNESS)	○	6.4.3
		CYLINDRICITY	⌭	6.4.4
FOR INDIVIDUAL OR RELATED FEATURES	PROFILE	PROFILE OF A LINE	⌒	6.5.2 (b)
		PROFILE OF A SURFACE	⌓	6.5.2 (a)
FOR RELATED FEATURES	ORIENTATION	ANGULARITY	∠	6.6.2
		PERPENDICULARITY	⊥	6.6.4
		PARALLELISM	//	6.6.3
	LOCATION	POSITION	⌖	5.2
		CONCENTRICITY	◎	5.11.3
	RUNOUT	CIRCULAR RUNOUT	↗*	6.7.2.1
		TOTAL RUNOUT	⌰*	6.7.2.2

*Arrowhead(s) may be filled in.

FIG. 68 GEOMETRIC CHARACTERISTIC SYMBOLS

Figure 6.1
Illustration of the use of American National Standards Institute (ANSI) symbols and their meaning from American National Standards Institute Manual, "ANSI y14.5M—1982. Courtesy of the American Society of Mechanical Engineers (ASME).

manager describes ANSI standards as a universal symbolic code, a set of symbols that constitute a literal visual language that can be used in molding plastic or machining metal:

Everybody, when you refer to ANSI standards, you set dimensional calls from ANSI standards. It can be readable from one person to another without verbal instructions as to how it's read. It's an industry standard used by engineering and manufacturing facilities and that's used world wide.

While this visual lexicon provides the language for conveying exactly what should be done in the shop, overuse of the symbols conveys a different symbolic meaning. One initial problem with the implementation of computer graphics and standards occurred because often the software did not contain the specialized symbols used by specific companies and building up the library of parts and symbols took time. While this is less of a problem today, a remaining stumbling block is that computer-assisted design was developed for commercial use and so its capabilities and library of parts do not always fit the rigorous standards used for military contract drawings.

Moreover, standards can be overused. This is not a minor problem of details but can contribute to large cost increases. Too much finely articulated information designated on a drawing by the overuse of standards symbols does not add clarity but rather signifies that highly rigorous production standards are demanded. The problem is that many engineers who work on the computer all the way from concept level to detail level do not understand the complex conventions of drafting standards and may either employ standards that contradict one another or use too many standards which can drive up the price on something as simple as a keyboard key cap, hardly a precision instrument. The drafting and CAD supervisor at Orion points out what can happen when overkill occurs in the application of standards:

We had this, a design engineer, whose main function was to design the key caps for our Orion keyboards. Um, some of the applications he used, he would use two positions. . . . He went ahead and used profiles and surface geometric callouts, perpendicularisms, parallelisms, and it was just totally overkill. And the vendor said "I really can't build this according to what you want me to do." . . . So it can really create an overkill situation and a really confusing situation. So we had to go back and tear apart all those key cap drawings and define 'em a little bit easier so they could understand them. And yet [it was] what this engineer thought he really wanted to get. And so it can be really detrimental to your process. And if you use overkill and if you use it abusively you misunderstand what they're all about. . . . Anyway, so we redefined it, um, where the vendor could really read it and make the tools for it and not charge an astronomical fee.

He also draws a correlation between the proliferation of specifications, which gives vendors an excuse to raise the price, and the high cost of military contracts:

It [the use of standards] can have a really adverse effect on what you want to accomplish. A lot of vendors will look at that and say, "Well, this is a real good opportunity to jack the price up of this tooling because he wants us to almost create the impossible. We'll let them know that, hey, we can handle that," and vendors will do this to you. . . . It [standards] can be very, very good, and they can be very bad for an organization. The military loves them. The military loves everything that's standard, and you deal with mil. standards, you deal with astronomical costs.

Clearly, the use of very rigorous and highly articulated standards on a set of plans signifies not better communication but the demand for rigorous point-by-point accuracy in the shop followed up by rigorous point-by-point checking to verify that the standards were met on the final product. Both of these procedures add greatly to production costs as did the redrafting of the erroneous drawing. Hence an engineer's noncomprehension of the symbolic rather than the literal meaning of standards on a drawing can be costly in numerous ways.

The example above illustrates that standards share with aesthetics a symbolic role as well as a functional one. The erosion of aesthetic elements through computer-graphics implementation has meant some loss of individual and company recognizability in terms of style. In some cases individuals try to compensate for this loss of recognition by overdetailing their drawings in an attempt to regain acknowledgment for their skill. This creates the problem of drawings that cost too many workhours for the function they are intended to serve. On the other hand, when an engineer overworks a drawing in the use of standards, the cost result can be exponential. It is a matter not merely of an overpriced drawing but of a geometrically larger increase in production costs. Once again, multiple levels of reading the drawing can introduce problems. In the case described above, the vendor does a dual reading of the drawing, reading both the general information in the universal code and also the embedded code—which signifies, by extensive use of standards, that a high level of precision is required, that a high level of skill and checking is demanded, and consequently that a very high price can be charged. Obviously, the socialized norm is that standards used in drawings must be applied in some sort of middle range between insufficient production information and overkill that drives up prices. But the engineer missed this socialization. This particular example is further tied to the restructuring problems raised by computer graphics in that the offending engineer was doing his own detail work from start to finish on the CAD system, without the assistance of an experienced designer or drafter who could have informed him of the consequences of standard overuse. He thought he understood the universal code of standards, although he made mistakes even there, according to the drafting supervisor. Even more dangerous was his lack of awareness of the symbolic code in standard use that could drive up production costs astronomically. The redrafting of his work averted the calamity but was nevertheless also costly.

Beyond such problems caused by the restructuring of individual work, implementation of computer graphics has restructured internal relationships among groups or departments within companies.

Workgroup Restructurings

Time Schedules
Before computerized modeling, the early phases of a large engineering project (such as those undertaken in civil engineering) were times when individuals could look at in-process sketches from various departments to gain information they needed for their own work. Now, three-dimensional modeling requires an individual or small group to construct an entire model before individual two-dimensional sketches can be obtained, and those whose work is contingent on other aspects of the design and layout must wait for the model to be completed before they can incorporate, question, or use information from it. This extends the time to completion for everyone involved.

Even when engineers use two-dimensional software rather than three-dimensional modeling, informants report that they receive incomplete CAD drawings, resulting in other "time-crunches" while missing information is gathered. To deal with such problems, drafters fill in the missing information, or companies attempt to streamline drawings. Unfortunately, streamlining through removal of some levels of detail appears to be a good idea but can lead to more problems given the indexical nature of visual information. Since drawings serve as conscription devices and boundary objects that mean different things to different individuals and groups, choosing which detail to include or delete can be hazardous. Streamlining also supposedly involves everyone who generates or uses the drawings, but virtually always those who

ultimately use the drawings in production are not involved in the negotiations. The Orion drafting supervisor acknowledged that his company
does not solicit input from the toolmakers, for whom the information
is ultimately aimed, especially if they are at a vendor site.

Because visual information employed by multiple users can have different meanings for different readers, those different readings further
index other tacit or visual knowledge that may not be explicit in a drawing but is part of the crucial arena of nonverbal knowledge that makes
up the action-centered skills of experience-based expertise. Removing
visual and written components from drawings can mean removing more
than surface marks on paper: it can mean the removal of a whole cognitive realm of information indexed to those marks for various users,
including those at the production level—potentially resulting in expo-
nentially expanding gaps of information.

Further complications generated by the removal of information from
drawings can arise because of the nature of tacit knowledge, compounded by the threatened position of drafters and designers. Because
managers and others know that the positions of designers and drafters
are threatened with the implementation of computer graphics, they may
feel that individuals or groups want to retain certain elements in drawings merely to justify their own jobs. Hence, the motives of those attempting to retain more complexity in drawings may be mistrusted
because of the job insecurity caused by the restructuring brought about
by computer-generated graphics.

The graphics manager from Orion pointed out that since engineers,
who are less concerned with the accurate transfer of information between groups as with the design itself, are now often in charge of the
whole drafting process on CAD terminals, they may leave gaps in the
information they provide on drawings. Drawings may not include
changes negotiated with the vendor or may lack other information required for updating other parts of the production process. Incoming
parts may be sent back or moved into storage instead of into production
because they do not match outdated drawings. The Orion drafting supervisor pointed out that this occurs because engineers do not regard
the drawings as important, once the molds for production have been
machined:

A toolmaker can have an engineer's preliminary design. A lot of that from there
on out is verbal communication and a lot of times is not reflected back into the
CAD file. . . . So we'll get these files, and, low and behold, we'll have a first article

piece that came out of that tool, the latest and greatest tool, and we'll use that sometimes to help us with the drawing because there's a lot of things that we notice that the engineer doesn't put in there. Well, low and behold, we'll find out that certain snap tabs have actually changed or that a location of a snap tab is no longer there. Or there's some kind of hole deleted or ribs or vents added that they're not reflected in that engineer's CAD file.

They wanna be able to say that the engineer signed this document off as being *true* to what we're supposed to be receiving and it doesn't look like the part I have in my hand. What do I do? And that has held up production. That has created a lot of, ah, chaotic instances out there in manufacturing. That has made a lot of parts sit on the shelf and be rejected. Or it's had people make a call for a deviation to accept these parts until we get good documentation, up to date. So it has a lot of impact, although its not visible in terms of engineering, the engineering world. . . . I feel sometimes that there're certain engineers—I assume they sit down at a CAD station. . . . I cringe because I think, "God, here we go again."

According to old convention, engineers did the rough and conceptual work, and drafters did the polished and detailed drawings. But today engineers do not yet do all the visual representations from start to finish. Engineers feel they have more important and interesting things to do than be concerned with the drafting minutia of documenting details and updating drawings. The Orion drafting supervisor attributes this reluctance to engineers' inability to recognize the necessity for drawings once the tooling has been completed and to their assessment of such follow-up work as lower-status work to which they can give less priority or attention. Nevertheless, if this lower-status but crucial update work is ignored, the information and relationships shared among groups working toward production are disrupted, and the design-to-production process is thrown off schedule.

Trusting the Social Network
Developing trust of a new kind of machine that has been introduced into a conventionalized work process is of course problematic—as is trusting the network in which it is enmeshed and for which it comes to stand. Because visual representations are also conscription devices, restructuring the work that goes into them means restructuring networks and then learning how to trust them. The issue of trust becomes dramatically complex when a company changes from using a three-dimensional plastic model to using computer modeling in engineering design. Those who use the final computerized model to generate drawings must trust more than the accuracy of the computer's calculations. They must trust

the software, the hardware, and the work of every person who has generated or entered information into the total system. Moreover, they must do this without being able to go to a plastic model they can study and literally measure. The Metrotech CAD manager pointed this out in the following statement:

What happens when you start making three-dimensional electronic models, supervisors and project engineers and project managers don't know how to go touch that stuff. And so you got these people working for you, now that you're sort of trusting them too, you know, like the piping designers that are now using this new tool. You used to be able to go and see how he was performing in his progress, and you have a hard time doing that now. It's hard to quantify progress on a three-dimensional model, three-dimensional *electronic* model because you don't work on drawings. Our old society was very drawing oriented. In a three-dimensional electronic model, you don't see one drawing on these pictures because you build the model. And when you're done building the model, if you want to make drawings after you build the model—okay, I'll make drawings. But to get to that stage right here by the old method, they would start making drawings of this structure. And now you don't think about a drawing. You start building a three-dimensional electronic model.

This CAD manager is making an excellent sociological point. As the computer is used to integrate all the information into one database, the group nature of the design process is highlighted. Users of the information cannot easily retain the myth that they work as individuals; they must recognize and trust the group and trust the machine that links them. As Becker (1982) shows in art worlds, so-called individual masterworks result from group processes that involve everyone from the canvas and paint makers to taste makers in the art community just as Callon (1986), Latour (1987), Law and Callon (1988), Bowker (1994), Bijker (1995), and others have shown the necessary networks for technological innovation and production. In civil engineering the piping designers, structural engineers, designers, architects, drafters, and computer operators—as well as the machine and its software—are needed to put out a new product. Coordination of these networks has been impeded by traditional distrust between white-collar engineering and blue-collar construction, as illustrated by the Metrotech CAD manager's remark:

Construction says, "Engineering doesn't know what the hell they're doing," and Engineering would say, "Construction are these overweight, stogie-smoking, get-down-to-the-brass-facts, don't-follow-plans-if-I-don't-have-to kind of people." And engineers always felt—not always—but the typical attitude might be, "It doesn't matter what I put on these drawing because Construction's going to build it the way they wanna build it anyway."

Despite such traditional boundaries, now both groups must trust the model makers as well. Computer hype and the vision of the great database in the sky in the following comment from the same manager suggest that CAD technology can do away with both such conflicts and the necessity of checking:

It [electronic three-dimensional modeling] has to evolve over time so everybody gets to understand the process and trust the process. . . . Now we can check the data. We can check less. . . . Our checking is less than it used to be. As we get to the database-in-the-sky kind of situation, everybody executes properly. You will probably see little to no checking. The machines can do most of the kind of cross-checking.

Given what the Orion manager, quoted earlier, pointed out about the problems he encountered with implementing the computer-assisted design system, and the need for human checking based on hands-on knowledge to make the computerized system work, this utopian vision looks far distant. The current use of the computer for graphics raises as many, maybe more, problems of coordination and trust as it solves. It adds other elements to the network of people and things that needs coordination, and it often disrupts systems of tacit knowledge around which groups could organize their activities. Such factors engender the need for more checking rather than less.

Restructuring Relations between Companies: The Question of Standardization

The multiplicity of proliferating restructurings caused through computer-graphics implementation reverberates beyond the confines of any one company and affects relations between companies as well. New methods raise issues of standardization in both record drawing procedures and software compatibility. One issue—the ease of entering as-built changes (those modifications implemented during final production)—into a computerized record drawing, which does not have the engineer's contract signature on it, had raised both practical and legal questions, as noted by the computer-graphics manager at a civil engineering company:

Now the issue becomes how do we issue "as-builts" or record drawings. One way to do would be to take those signed drawings, the hard copies, and physically make the changes to 'em. That loses the advantages of having a CAD system. The other way to do it would be—Well, we've got a bid set which is a record set, but we also know that the set that's over here in the computer is, is, a mirror image of those, and so we can go back in CAD and make changes to those. Now, the owner has two sets of drawings—one set that has the signature of the engineer that was the bidable documents and another set that's the record drawing. And so CAD has made that process kind of muddy as to what we do. And I've surveyed other companies and everybody does 'em differently. . . . Probably, what everybody's probably waiting for is a case like that to be settled in the courts or for the boards of professional registration to deal with it, which they have not yet done.

The suggestion is that social negotiation will have to control a problem generated by use of the machine—another example of the action of machines in social networks.

Incompatibility among software developed for computer graphics is a different sort of problem raising both the need for standardization and the need to maintain flexibility. The development of new software has facilitated transfer of data from one graphics system to another. However, in many cases content can be partially or completely obliterated or so badly distorted that it is unusable. Space does not permit me to quote the proliferation of crucial detail losses engendered by conversion software such as Initial Graphics Exchange Specification (IGES), which worked well for some systems but applied to others would not even retain dimensions. The IGES software, developed through a consortium of large manufacturers who used many outside vendors for their component parts in the aerospace and automobile industries, was the first step to try to solve the conversion problem but remained problematic. This was so, first, because information was lost in the transfer and, second, because if designers fudged and did not do the whole design in true geometry, as dictated by the system, lost information could not be regenerated.

These issues are larger than the loss or distortion of information. Since a particular group of companies contributed to the designing of the IGES software, it worked best on the CAD systems either belonging to those companies or used most widely by them. This, of course, meant that smaller companies who make graphics system products or design graphics systems could be squeezed out of the market unless their product or design was compatible with the systems being used by industrial giants that began to set the standards through such products. This is hardly a surprise. Noble (1977) points out exactly the same kind of dominance by large companies in the setting of all standards, even when the standards are set by congressional committees, which are usually made up of corporate representatives. The Orion manager further stated that the crucial need to be able to transfer data, coupled with the availability of only one transfer software (at the time of this interview), dictated what was deemed compatible. The IGES capabilities, therefore, dictated what graphics systems user companies were willing to buy, thus not only restricting the market but also setting the standards in a manner similar to the way IBM, Apple, and Microsoft have set the standards in the personal computer market.

Despite the numerous problems of incompatible systems, users do not universally support one standardized system. The Metrotech graphics manager pointed out the importance of retaining the flexibility of using many systems to avoid a standardization that threatens that flexibility:

All these companies went and did what they want to do. Because of that, right now we're flexible. I can call on any of these. Somebody [sic] want to talk about Autocad, the Intergraph conversions. They did it in Houston. I get on the phone. Somebody want to know how to use Computervision, they call me: I'll tell them. Somebody [sic] want to know how to use Calnet. . . . We have a tremendous amount of diversification. We do because all these guys went and did their own thing. . . .

It's good, and it's bad. What is bad about it? I said, "We did this fifty times. Just think. You should have done it once." I mean it's inefficient. And yet it's good because we need to do that in each of the companies to make sure we— we're going down paths right. . . . I disagree about lots of things about standardization. I agree with some things about standardization. Our company is banking heavily on the payback of standardization. By coming together with one common system we're gonna be, we're gonna end up being structured to where we're gonna have our system. And the only way we're gonna deviate from that system is if someone comes in and says, "I don't like your system, I want my system." And then we're gonna have to do something. And more than likely, what they preach to me now is, if a client comes to you and says that, "The price for our system is going to be one dollar. The price for us to customize to their system is going to be ten dollars. If they don't like it, leave." I mean, that's sort of the way it's being put to me. I think it's bellyhoop. I think when a client comes in and says, "I want this, and you're competing with Floor and Brown and Root, and see up front, and Stone and Webster, etcetera, etcetera" for this job, I think we're gonna be—. They'll say, "Shit" and we'll say, "which corner," kind of thing. So I'm a little leery of standardization in that it's gonna limit us. And a lot of the flexibility that we've had up to now has been an asset, and we may not have as much flexibility in the future.

Despite his faith in the new technology and standardization—not unlike the misplaced faith of management in Taylorism earlier this century— this CAD manager also sees its shortcomings in terms of flexibility in serving diverse clients.

Government agencies, just as commercial industries, experience problems with the incompatibility of computer-graphics systems used by the various companies with which they contract and have been taking steps toward setting standards. However, even their first steps have been problematic because of varying needs for flexibility, depending on the agency and its applications. Indeed, problems encountered in the very attempts to standardize point up the great variability in uses and needs in the graphics applications of different organizations.

Despite the inability to agree on a universal standard and a desire to retain flexibility that reiterates the point I have been making about the resilience and necessity of mixed practices, there seems to be an almost magical belief in what the Metrotech graphics manager calls "the great database in the sky." He pointed out that his company had done numerous studies to look at designing and documentation process in order to streamline it but that, in his words, "Every time, someone worried about their own interest," so that a more standardized format did not arise. I suspect such "worries" were actions taken to protect messy situated practices that ensure that the work actually gets done. He continued to maintain that "we've got to get our act together" and "find one best method." This kind of rhetoric still permeates discussion with many CAD managers even while they are complaining about the problems generated in the restructuring of work and work relations that implementation of graphics systems engenders.

Conclusion

In many instances, when a complete drawing or drafting job is handed over to engineers to do everything from concept to detail, drafters and drafting departments are filling in the gaps in engineers' work to ensure that the job is done in a way that produces useable drawings. The data presented here reveal patterns I also encountered in the case studies where I found that attempts to circumvent or undermine some uses of CAD were not merely a form of Luddite resistance to mechanization or merely a stubborn resistance to changes occurring in the labor force. I saw reasons to distrust the prevalent view that younger designers, drafters, and engineers will more fully embrace the new technology as they replace senior people. Everyone from senior drafter to junior engineer faced restructuring problems raised by the new technology and had to find ways to reduce their negative consequences for design. What has been viewed in the past by management as resistance can be seen here as efforts to overcome the restructuring problems introduced by CAD. Redrafting from first-article products, redrafting to correct standards, and inventing new ways of fudging are just the beginning of the mixed practices that people use to get the job done using new computer-graphics technology. Technology has a role to play, and engineers, drafters, and designers are glad to have it because it saves time and effort in repetitive work. But one single method is not right for all circumstances. Just as the drafters pitch in to clarify engineers' insufficient drawings, creative people using the new technology find other ways to mix paper and electronic practices so they can retain familiar practices and cognitive spaces necessary to do their work and still take advantage of the expediencies of computer graphics. We can now turn attention to those mixed practices.

7

Mixed-Use Practices: Combining the Electronic World and the Paper World

A chip designer at a highly visible computer manufacturer switches from the monitor screen to paper copies of his design in order to analyze his work. The manager of the graphics department in an aerospace components firm complains that his designers use more paper than ever now that they do all their designs on the computer: they print out fresh copies for every consultation. Engineers, designers, and drafters from companies that make cookie factories and nuclear power plants to those that design micro chips and satellite communication systems use computer-graphics systems but also still use pencil, paper, and pen. These are mixed practices. This chapter addresses the key question raised by the restructurings discussed in chapter 6: How do people cope with the changes to get their job done? Not only can computer-graphics technology not provide one best way to accomplish the visual communication of engineering design, but computer graphics must be and are being helped out by resourceful patching in from the older paper system. Moreover, people employ a variety of mixed-use practices that fit the particular circumstances of their work to get things done accurately and on time.

Situated Knowledge and Mixed-Use Practice

The use of computer-graphics technology at different industrial sites is context specific. The manner in which computer-graphics literacy is used at one company is not necessarily the manner in which it is used at another, nor is one approach the best use of the technology at all sites. This is not universally recognized.

One-way deterministic assumptions about how computers are used in design mirror earlier ideas about the relationship of literacy to development. Using substantive field work, Scribner and Cole (1981) have countered simple notions about the effects of literacy, documenting that print literacy in and of itself has little impact on development. The effect is use dependent or context specific. In their work with the Vai, these authors have noted that members of this west African culture may achieve literacy in three different languages and use each language for a specific function. Proficiency does not cross over from one language to another, but rather each is practice oriented to the particular use in which it is employed. Similar patterns apply to computer use. Many managers feel that computer graphics can be applied in a simple and unitary fashion—the kind of the mythical linear fashion presented to them in computer sales force hype. They are sold a push-button fantasy world of robotics that will radically cut personnel costs. In the actual world of messy practices, even the actions of incorporating a highly rationalized technology such as computer graphics is more variable and more interesting than such mythology would suggest. Just as the use of written literacy varies by context for the Vai, the use of electronic graphics literacy varies by context for designers.

Influence of the Paper World

I begin by describing two types of site-specific, varied practices. *Mixed-use practice* both uses paper and electronic modes for representation, communication, and analysis. *Differential-use practice* uses various electronic options available on site and distinguishes one or another for specific tasks. Often the differential choice of specific software or hardware for specific tasks may emulate older distinctions inherited from paper-world practices. For example, rough sketching for note taking and analysis would use lower-quality materials, and finalized drawings would traditionally use higher-quality materials. Another differentiation might be the last ghost of older divisions of labor between engineers and drafters. Such varied use protects the work and the worker from the regimentation of all-too-linear computer dictates and suggests that a completely paperless world for design engineering is not only a futuristic myth but also potentially unproductive and dehumanizing. Human creativity can be enhanced, rather than replaced by, technology, and designers can use creative choice, mixed practices, and variation in the applications of a new technology to produce creative work.

Of approximately thirty industrial design sites I observed, every company was using computer graphics in some manner—but not necessarily the same manner. No one best way to do design work—the perfect linear process—has been found. Through the resilience of creative thought and practice, designers, drafters, engineers, and managers are making compromises that work. These compromised practices are not flawless, but then no work ever is. Design work is made up of the messy daily practices in a world where interaction between people using conscription devices and boundary objects patches up the recurring errors and misunderstandings that are characteristic of the ambiguity of human interaction. To call these approaches *strategies* would not be accurate because that would suggest calculated forethought. While the compromises function like strategies, in that they facilitate the achievement of company-oriented job goals and individual-oriented work goals, the process of their evolution is much more mundane and interactive. They are practices constructed by individuals in the situated context of doing their own work or constructed by groups in the process of team interaction. Mixed and differential use of paper and electronic options solves some of the problems raised by the introduction of computerized graphic systems, whether the problems are more purely technical in nature or are oriented to the work itself and the relations of the workers to one another and to the technology.

Mixed-use practices are not merely a transitional phase to a completely electronic system (though they can also be that in some cases for some, but not all, practices). Unfortunately, in the hype of computer sales brochures and of some graphics systems managers, such variation is seen as a problem rather than a solution. Recall the comments of the Selco turbine package manager who saw his team's use of CAD as "an expensive, fast eraser" as "wrong." Indeed, mixed-use and differential-use practices are extremely important in that they help maintain flexibility in the system. Boundary objects and conscription devices are kept intact or allowed to evolve through these practices. Tacit and visual skills, indexed by visual representations, are preserved and nurtured rather than choked. Once we drop the assumption that there is no one right way to use an electronic graphics system, it is easier to see that if a company finds that record keeping is what its computerized graphics system does best, then such use is still a paper storage and documentation problem solved even if the company's best designers avoid it. Designers probably have sound reasons for any resistance to a complete changeover to a paperless system. Understanding and discussing some of those reasons are part of the aim of this chapter.

Mixed-Use Practice: Moving between Paper and Computer

As the case studies illustrate, a good portion of design engineering knowledge and communication is visual in format. Engineers, designers, drafters, and their managers at the comparative sites reiterated this in their comments about design work. At a company that builds aircraft control panels, a designer illustrated this preference for visual communication by walking to the chalkboard and drawing as he answered my questions about the role of visual communication in his work:

My instinct is to go right to the board. I'm very graphic oriented. I can't talk more than ten minutes without I start drawing pictures when we're talking about the things that I do. Even if I'm talking sports, I invariably start diagramming what's going on. I feel very comfortable with it or find it very effective.

This designer, like the newly promoted engineer at Selco who fought to get her drafting board back, is pointing out his dependence on the visual process of drawing both to communicate and to think out the initial design. He also states that this visual and manual thought process of drawing precedes the formulation of the written specifications for the project. Like the Selco designers and those at other sites, he emphasized the importance of drawing processes to work out ideas.

Paper for the Quick Capture of Initial Ideas

A primary situation in which designers return to paper, whether it is a hard-copy printout or a hand-drawn original, is during analysis phases. This may be very early, during the conceptual work; midstream, to keep sight of the whole of a project; or at the very end of a design cycle to check the very last adjustments to a mechanical design to test a chip, or to check the drawing itself to ensure that it includes all the requisite production information and meets the company standards. An Orion Computers chip designer gave this example of his use of paper for initial conceptual work:

Sometimes I just sit at my desk and think and do mind images. . . . That's a pretty efficient way for me to start. Then I often write down equations or diagrams on paper. Paper's real convenient. . . . Usually when you're using a CAD station, you're running a piece of software, . . . so sometimes I use paper when I'm sitting right there at the CAD station, and I'll do some little drawings right there in front of me. And then once I think I have something of value, I put it in the CAD station. I'll also use the Mac to do drawings and sometimes write out dots and equations, things like that.

Both I'll write, I'll type thoughts down, or else I'll draw pictures or something like that. There are a lot of times when I start on a new section of my task, I may go to the Mac and do drawing, do very high-level things and kind of play around with things, with how I might want to approach things. . . . In most cases it's easier for me to do it on paper; it's faster, and that's the key. . . . It's easier for me. Once I've had some kind of a big idea of what I want to do, I narrow it down on paper, and I enter into the CAD system.

The drafting supervisor at the same site corroborated that such practice was common:

A lot of the industrial design[ers], they'll use CAD to a point and do a 3D solid with it, um, and a lot of time they start off with sketches, um, and then the concept comes over into a CAD format.

A second chip designer at Commutech, a firm that designs and manufactures devices to protect cable television signals, described his somewhat unconscious choice to use paper instead of a Macintosh work station to get his ideas down as quickly as possible:

There were times where you just had a quick idea. I'd be sitting at my desk and have a quick idea that . . . can help you get down something, just anything quickly down on paper—sort of test it out—to see it in visual form because there are too many parts to sort of keep in your head. Then, [I] didn't want to go upstairs and mess around with the machine. [The work] may be not final and pretty. So at times it was just quicker to get a little something down on paper even though you might throw it away. In fact, because you would probably end up throwing it away. There's this feeling, okay, that machine is for my final design.

What this designer is pointing out is the residue of the paper system in that he equates the computer-generated document with a final draft, which in the paper world would have been rendered on better-quality paper using professional-quality drawing materials that are too expensive to use for quick sketches. Such sketches could be jotted down with any ordinary pen or pencil since they will be thrown away later. Supporting this analogy is the fact that the computer terminals are expensive and must be shared, just as art materials used for a final draft are expensive in comparison to ordinary paper, pens, and pencils.

Getting the Whole Picture
The same chip designer added another point that both mechanical and electrical designers often mentioned. Looking at one or more paper copies of their design allowed them to get a more comprehensive view of the design than the monitor could provide since it could display only one layer at a time or a very small total view on the screen:

One of the reasons for that is because when you have a hard copy you have the whole thing here. . . . On the screen it tends to be one sheet. The Macintosh has the facility for opening up all the sheets of the schematics, but they're all back behind one another. Well, these [paper copies] are, too, but I can be holding two or three of them here at one time if I want to.

A chip designer at a well-known Silicon Valley computer firm similarly described his overall work habits as a mix of paper and screen use. He also pointed out that when he is ready to analyze a drawing, he uses the paper copy to "get all the pieces" before he does a final print. He described the screen representation as "like a jigsaw puzzle" in which it is necessary to see the whole to understand the parts:

The only problem is the design is so large the screen is not large enough. It is easier to look at the sheet than the tube with only one section [displayed]. I have to remember more as I move from one section to the next. I have to remember what was on another page. In my mind I have to remember what was visually there before. Before [on paper] I could scan over the page. Sometimes I have to print it out to put all the pieces together. There is a limit to what I can remember, one page to the next, especially if I go two pages or in a complex design. I have to have both pages.

For understanding the total picture—how parts relate to one another— this very proficient CAD user, who notes he can draw on-screen as fast as he can type at a good speed, finds it is necessary to mix paper and electronic materials.

The same engineer also noted the limitations of screen work when a team is involved, such as in system design. He points out that on such projects ordinarily three or four engineers would gather round a large drawing and "draw on the board what they think and everyone talks to that." He notes that with the CAD system it is not as easy to communicate with others, again because the small size of the screen makes it hard to see both the total picture and the minute details: "When there are more than two people" looking at the screen, it is "difficult for some reason," and then people seem to "go to the hard copy." Asked if it was confusing on the screen, he replied, "I think it is, especially if the person is not familiar with that design." He further noted that since he had done the design, he knew where on each page the information was stored. Given the problems he described of relating information on various pages, it would be even more confusing to someone less familiar with the relation-ships of the pages to one another, especially if no overall view was pro-vided. An analogy might be trying to find your way in an unfamiliar city by looking at one page of a map book in which each page contains only a small portion of the whole, without reference to the overall map that shows the relationship of the pages to one another.

Here we see an overemphasis on the site-specific nature of the display of separate detail pages, an overall lack of flexibility of the larger boundary object in accessing multiple readings by other participants, and hence a rather forced isolation. However, the resourceful engineer patches up the system by turning to the paper copy, or several of them, to fill in the missing information and to reconstruct the boundary object and conscription device. He can keep in view how the details relate to the whole as well as orient collaborators in order to keep communication open.

"Hard Copies" for Analysis

Connected to the need for a total view is another reason for a return to a paper copy. Again emulating the paper world, many designers reported that they used paper printouts to analyze and make adjustments to their design, which they would later incorporate back into the electronic record, after the problem was worked out. This is another example of how designers use sketching to try out ideas and to think with during analytical phases of interaction. They mark up the computer-assisted design in its hard copy format, and treat the electronic copy like a final draft. Looking through his collection of rough-draft, red-lined schematics, the Commutech electronics engineer described his analytic practice on a job he had just finished:

I already had everything on the Macintosh by then. . . . Even then there was a combination. I was working with pencil and paper there, too, so I moved back and forth. . . . Somewhere—I realized that I—it wasn't right, and I had to add this guy in here. But I think I did a lot of sketching on the original paper with this here. And I started drawing this stuff in. . . . Which is very typical. You know, you red line a drawing like this and then enter it. Again, if this were pencil and paper, I have to fill in, redraw this, but I have to redraw the whole sheet. . . . Or erase this, and [it gets] real smudgy. That way it's on the machine now. I can print out this sheet and I can kind of mess around with pencil or red ink or whatever. And—and so I'm doing [it] in pencil but I'm doing it on the sheet that's on—from the machine and when I see my corrections. Now that's all there together, and I can look and sit and examine it. Now is that correct? And several times it wasn't. So I would maybe scratch more, or I might get another printout or a sheet.

Bucciarelli (1994), in a finely detailed account of engineering design, notes similar sketching practices prior to entering designs into CAD systems and the hand marking of changes on formal hard copies. In another instance, the graphics system manager at the aerospace components firm quoted at the beginning of this chapter noted that his designers used paper versions for analysis. "Any time they have a meeting, they need paper." He pointed out that designers and drafters at his site also use paper or hard copies "when it goes to check"—meaning when the drafting is inspected for accuracy and clarity. He complained that they "use more paper on the computer" than they ever did on the drafting boards. The intermediate hard copies are not kept, so every time the drawings go "to check," both the old and the new ones are needed to compare the changes. Part of the reason for the excessive preoccupation with checking in this industry is that designers are working to military specifications or standards (discussed in the previous chapter), which are much more extensive than those used in commercial industry and apply not only to the project being designed but to the manner in which the drawings themselves are done. The fact that the combination of very detailed standards and computer graphics leads to the use of more rather than less paper points up the importance of mixed practices when meticulous work is being done.

Differential-Use Practices: Hardware and Software Choices

I have defined differential-use practice as making choices among various electronic options, such as available hardware and software, and distinguishing one or another for specific tasks. Such choices may echo older distinctions from paper-world practices such as selection of less-expensive media for rough sketching and more-expensive media for final drawings, or they may be the shadow of older divisions of labor between engineers and drafters.

A common type of differential use of computer graphics that was practiced at several sites was using one type of hardware or software for one phase of the designing process and using a different one for other phases. One of the most prevalent practices was the use of Macintosh workstations for earlier, still-conceptual work and the use of CAD workstations for fully developed, detailed work. In some cases, this division of electronic labor mirrored prior divisions of labor between engineers and drafters in that engineers would do early conceptual work, and drafters would finish the precise detailing that clearly communicated to shop workers. In a few cases, the same engineer would do both conceptual and detail work, only a ghost of the prior division of skilled labor remaining as a distinction in choice of software. The important point here, however, is a distinction in work practices that once again emphasizes the importance of visual process in working out ideas in the early stages and the necessity of a medium—paper or the more accessible Macintosh workstation—to do visual note taking and outlining, the paramount first steps in design work. The chip designer at Orion Computers makes such a point regarding his own work:

I guess at the beginning . . . of a project, I'll use the Mac more because I'll be writing thoughts down. . . . I'm not doing the design part. I'm doing the specification and the very high-level conceptualization. What do we want to *do*? I'll use the Mac for very high-level stuff, maybe.

Both I'll write, I'll type thoughts down, or else I'll draw a picture or something like that. There are a lot of times when I start on a new section of my task, I may go to the Mac and do drawing—do very high-level things and kind of play around with things, with how I might want to approach things. . . . In most cases its easier for me to do it on paper. It's faster, and that's the key.

Getting Ideas Down Quickly and Cheaply

The graphics manager at Metrotech civil engineering mentioned that management at his company encourages engineers to use Macintosh workstations for early conceptual work because they are less expensive and not because they are more applicable:

This company, here, believes that we should use Macintoshes for a lot of conceptual design because it is purely drafting functions. You're not doing any detail-sy stuff. So you're doing, say, simple drafting functions, and you want the cheapest way and easiest to learn because you're gonna have ten people thrown at the job real quick, and you want them to come with their tool box, and you want them to go to work immediately.

Expense factors are only a small part of the issue. A systems designer in satellite communications pointed out that CAD systems are available at her company's site but that Macintoshes are preferred for much conceptual work. In this case, it is not the engineers but the technicians who are more proficient with the technology and who redraw early versions of a design so a clean copy of an engineer's ideas or sketch is available for collaboration with others. She describes the process:

There's a whole . . . group of people: that's all they do, is draw Macintosh pictures. That's their job. . . . I'll draw it up, and then I'll give it to this person to make it look nice, to make it look something presentable. . . . They're in the data processing group. . . . They tend to not be engineers. . . . They tend to not have a college background. And they just draw pictures all day. . . . I like to use it occasionally, but it seems to take a lot more time. If I can get a few pictures and have somebody drawing them for me, it just seems to take less time. . . .

So they draw these functional pictures and that tends to be Macintosh implemented, and then when the designs really start getting detailed, yes, they start going to CAD. But for proposals they tend not to be CAD designs yet. Only in mechanical-type things will they already jump into a CAD-type of approach, but usually that comes later.

She is not only describing a differential use of hardware—Macintoshes and CAD workstations—but also pointing out mixed use in that the engineers often draw hand sketches that the technicians render on the Macintosh in a clean format for proposals, echoing the former division of labor between engineers and drafters. Yet even this is varied in that sometimes the engineers redraw their own sketches on the Macintosh, while other times they hand it over to the technicians. She goes on to point out that one of the reasons she hands over her work is because she is an infrequent user of the Macintosh and therefore is not always up on the latest updates to the software. She adds:

But anyway, it's always a trade-off. If you have too many things to change or too much input, sometimes it's easier to draw yourself because it's a little confusing for them sometimes. . . . Lately engineers aren't drawing that much. . . . To be honest, I've gotten the attitude now that somebody else always works with that Macintosh so much better than I that I very rarely do it unless it was a Saturday or Sunday and we were desperate. . . . I always draw them by hand, I think, and then I go and put them into the keyboard or have someone draw it.

[On a previous project] they brought in, like, three or four Macintoshes, and we all sat there and used them for six months. Yeah. And you get real quick. . . . We call them Mac-gods. You become a Mac-god (laughs) right away. . . . But if you use it so much, you get to where—oh, you just do all these things. And so we all became Mac-gods for, ah—demi-gods—for six months (laughs) and then we lost it.

This site spawns Mac-gods among its technicians and engineers who use Macintoshes frequently, but when they become infrequent users, they return to the realm of mere mortals. Choices between workstations, like choices between the computer and paper, often are made on the basis of quick accessibility.

Accessibility versus Power for Different Functions

At a site that makes radar systems for the navy, chip designers differentiate between software, rather than hardware, depending on application. They use one software for chip design and another for board designs.

My informant, an electrical design engineer, noted that he used to start his work on paper four or five years ago and that although some people in his section still prefer to work on paper, he has "gotten comfortable using CAD." He stated that designers in his group prefer the software with "more powerful options" for chip design and the software that is "more accessible" for conceptual work and board designs. This was not a planned strategy but rather the outcome of structural change in the company. The software was purchased by two different managers in two different departments who, as my informant described it, each made his own decisions and did not talk to one another. Later the two departments merged, and design engineers now have access to both sets of software.

Multiple Mixtures: Two Different Software Choices Plus Paper

The choice of one software or hardware over another for conceptual work is similar to the paper-world practice of early sketching as a form of visual note taking and using better materials to create a clean, final draft. In fact, the designer above characterized the accessibility of the software he uses for conceptual work in terms of pencil and paper: "I like to use software that's easy to use or I might as well do it on paper." Using accessible software here does not result in a reduction in drawing time because the work must be redrawn onto the more sophisticated system since the two systems are not compatible. My informant said that he used his conceptual drawing as a guide line for detail design that would be done later. He also pointed out that after the conceptual portion was down, it was then used to write a specification, illustrating the visual essence of the design work in that the visual preceded the verbal articulation of the requirements. His overall work habits are a mix of paper and screen.

In the Selco turbine package study, Sharon, a design engineer, noted that when she started laying things out, she came to really understand a design problem and was able to analyze it, a parallel in the visual realm to one accomplishment of written literacy. Similarly, James, the designer whose work practices I have been discussing, noted that when he arrives at an analysis stage in which he will be making changes to his computer-generated drawings, he turns to the paper copy: "I come up with the idea and put it on the screen and then pencil changes." He points out that if there are "two ways to do the same thing," it is "easy to copy into another drawing" because he "doesn't have to redraw." He describes his process:

Once I've done the first one and think I've thought it through, that's my first cut. I look at the hard copy. Very rarely, the first [copy] is what I want. I find things missing, things I hadn't thought about. . . . If that picture gets very large, it's easier to take a large piece of paper and draw it. Just about everything else [I] can enter without paper first. When I first got interested, I had never used CAD before. The first thing I noticed was I could not get the whole thing. I could shrink it, but then I can't read it. I really wanted a huge screen about this big [arms spread to full span]. But then I really liked it.

But even given problems such as scale, which he mentions above, James pointed out that it would take him more time to draw by hand and write specifications and then copy everything onto CAD. Instead, he prefers to "move onto the system early," even if he has to recopy from one software to another. He modifies the trade-off of some loss of the total picture by keeping paper copies nearby for reference in exchange for more flexibility with varying iterations and lessened, if not total, removal of repetitive redrawing chores.

One of the chip designers further noted another limitation of the restricted size of the computer screen in reference to the kind of scanning one does of a visual representation when looking for something in particular. He stated that because of the "restricted view you have to move the window. . . . The shapes are hard to find. You have to look at each section and especially those that are small would be hard to find." He also pointed out that working between the full level and detail level on the computer requires a time lag because it "takes the computer time to draw all that." So therefore "after the detail level is pretty well done, I keep a full hard copy next to the computer. I'll use the hard copy along with the computer to make changes." Again, the patching in of the resourceful designer covers the inadequacies of the system by turning to paper copies.

Differential Practices Onscreen: Autorouters and Manual Routing

Understanding the importance and success of these mixed and differential practices involves looking at the work product—that is, the drawings themselves. These visual representations, along with what informants have said in the previous chapters about the process of making them, reveal some of the issues that link the social organization of the work and its product with the social organization of the workers. I have already discussed in chapter 5 how computer graphics affects aesthetics and standards in drawings. An example of a mixed-use practice that addresses some of the computerized constraints on aesthetics through a medium that allows for some malleability is the use of interactive autorouter software and the manner in which turning back to manual routing, still done on the computer screen, allows for better design work.

Autorouters, as the name implies, automatically fill in wiring between component parts in an electronic design. A chip designer at Orion pointed out that at its best autorouting is a mixed practice of manual routing accomplished through interaction with the computer screen and automated routing accomplished by the computer. He noted that the newer autorouter software and hardware lets the designer choose which portions of work to allot to the computer and which to work out themselves:

Most of the routing's done by hand. Inevitably, you have to go in there and change some of it. . . . There's some things the autorouter can't do. . . . Autorouters aren't as smart as I am. They just do things much faster, infinitely faster. So for the vast majority—and that trade-off is worth it, you get your chip out a year ahead of time—autorouters do most things. But a lot of times, ah—chips are made up of what's called I-O [input-output] pads. They're the parts around the chip that connect to the pin, and they have special things in them. They are the part that connects to the outside world and must protect against all the evils in the outside world. And a lot of times the routing of the inside core and the I-O stuff isn't very well done. So you do it by hand.

He elaborated further:

The computer also places all the transistors inside the chip, and sometimes I have to move them because I don't like the way it placed them. Or I may have some constraint that I cannot explain to the computer, so I'll let the computer do what it can and I'll adjust [the design].

As other designers—both of computer chips and of civil engineering plans—have pointed out, even in using graphic systems, elements of personal style still show through. The same Orion designer points out the distinctions that can be seen in the final computer-generated design revealing which portions were done *by* the machine and which were done manually *on* the machine:

Depends on the type of routing . . . but for the most part computer-routed chips are much more regular . . . in their structure so you'll see these real regular patterns. It—the routing is the connection between different elements. Sometimes you'll have, like, a RAM. A RAM is like a million of the same structure over and over and over and over again. So that looks very regular, and it's kind of pretty. It's almost art. It is art. It's really pretty. And whether that's done by hand or done by computer, it'll look the same. But routing between different— different types, types of things, if done by hand, it won't look nearly as regular or, ah, structured. . . . More gridlike. They route based on grids, and they place things on grids. But it won't be as efficient as hand routing. . . . Because all they understand are grids, so they place things on them. They come up with the best method of placing your elements on grids and wiring them together. But grids aren't necessarily the best way to do it. . . . A good layout person will always do a better job than a computer. It takes them forever. You don't have time.

He goes on to make the distinction that the autorouter is good for general, uncomplicated things—the general case—since it is, as he says, programmed with an algorithm for the "lowest common denominator," which always attempts to make everything in the design into a square format. At the same time the designer also has opportunities to change what the computer has done, allowing for another type of differential practice:

The computer has an algorithm programmed into it by somebody. . . . Those algorithms are usually the lowest common denominator, like, so they're for the general case—what would be the best way to do something. And inevitably that's not the best for my particular chip. So I'll have to make adjustments for what they do.

And they provide you fantastic methods for changing what they do, um, and for estimating the value of that. . . . The whole process of routing and placing the stuff can be interactive, okay. So you can just tell the computer to go do it. And there are a number of different stages in order to really finish the whole thing. . . . Between each stage you can make changes. You can undo parts, ah, do whatever you want.

So you look at the way it's placed things. Usually they try to make a square out of all the things that they place. But there might be a piece in the top left corner that you actually might *want* to be in the bottom right corner, so you'll move it. You know, they just have an algorithm that says, "I need to make a square because chips are better if they're square." And they don't necessarily know where everything goes. They don't. Part of the reason they don't know where everything goes is because they haven't routed it yet. Once you route it, you find out where all the connections are, so you spend a lot of time iterating on routing it and placing it and moving things and, um, fixing things in certain locations. It's tedious, but I like it. (laughs)

This design engineer is actually pointing out the interactive nature not only of the software but also of the act of designing itself, in which some elements must remain flexible until other elements are added that will have an impact on the placement of the other parts. He is also describing a mixed practice that is a combination of the automatic features of the technology and the manual, visual, and experiential skills of the designer, also know as tacit knowledge.

A final example illustrates the willingness of designers and drafters to use their experiential knowledge to tinker in innovative ways with mixed-use and differential-use practices in the face of pressure to adapt to new technologies and meet regular work goals and deadlines. Thus far, the examples provided have been of the use of computer-assisted design only—the CAD part of CAD-CAM—because that was the only part my informants were using. In this case, designers employed the CAM (computer-assisted manufacturing) technology to solve problems introduced by the CAD portion of the technology.

Designers and drafters at a well-known computer firm were working on a new monitor casing design. They were having a particular problem matching the corners that came together at the screen edge because they were recessed, concave, and curved simultaneously. As discussed in chapter 7, in paper-world practice such complex curves would have been rendered by fudging—that is, by using a malleable template to trace the line without computing all the points along it. But that option is no longer possible in computer-assisted design, which dictates that all the points must be input. So designers developed a new way to fudge, by tinkering with the technology. Using their approximation of the points on the curve, they had the machining function (CAM) produce a foam iteration of the part, examined it to see how far out of alignment the adjoining sections were, adjusted the points, and repeated the process, over and over, until the corners matched up. Such use is hardly what the technology was designed for. Indeed, it is the very opposite of the clean, linear efficiency purported by computerization. However, as we saw in the tinkering between sketches and plastic iterations in the design of the Optimed Inserter medical instrument, such tinkering is common to engineering design work. Here the CAM part of the technology has simply been enlisted into preexisting practice in order to work around an otherwise almost insurmountable problem, introduced by the CAD part of the technology. The problem was solved and work goals and deadlines were met.

Problems Solved by Mixed-Use Practices

Those who turn to mixed practices—to tinkering, to the hard-copy paper versions of designs—work around the shortcomings introduced by computer-graphics systems. The use of mixed hand-sketched and

computer-generated drawings, and even CAM-produced foam iterations, retrieves some of the physical sentience, action dependence, and
context dependence that Zuboff (1988) says are disassociated from
knowledge and practice in automated industry. As Lave (1988) points
out, real knowledge is rooted in the situational practice of everyday life
and work; the mixed practices allow designers to integrate their situated,
traditional skills and new technologies.

While some CAD advocates might assume that users of mixed practices
are infrequent users of electronic drafting tools and hence do not take
full advantage of computer systems, this is not the case. These designers
are bivisual (or perhaps even multivisual, given the number of visual
codes they must manipulate on various machines, software, and on paper). They work between the visual languages of paper and electronic
visual representation. Like multilingual speakers who "code switch" in
specific domains—such as native language for gossip and second language for business—these multivisual users choose which mode is best
for a given situation, such as paper and pencil for capturing ideas, hard
copy and red ink media for certain types of analysis, and computer
screen and paper printouts for keeping total views in sight or for making
corrections. They simultaneously use electronic media for other types
of analysis and final, corrected, and updated versions of their work, inge-
niously drawing on the strengths of all the media available.

Similarly, makers of visual representations who use computerized
graphic systems also make distinctions between the electronic media
they use, differentiating which hardware or software of those available
on site are better for a given application. Macintosh terminals and software may be chosen for early conceptual work, and CAD workstations
may be used for detailed renderings, or one software on a CAD system
is chosen for rough drafts and another for final ones.

All these illustrations point to the resilience and creativity of engineers
and engineering designers, drafters, and management. Computers cannot be brought into design work according to one best method, even
within a given company or for a single application. Individuals and teams
of designers make computers useful when they find ways to fill the gap-
ing holes left when CAD systems are implemented to replace hand drafting. A joke that made the rounds at computer conventions captures the
estimation of engineers and what the computer industry has promised:

A computer salesman dies and meets St. Peter at the Pearly Gates. St. Peter says,
"Where do you want to go—heaven or hell?" The salesman expresses his sur-

prise and confusion that he is allowed to choose, so St. Peter says, "Would you like to have a look?" First, he shows the salesman a videotape of heaven. People are strolling around through the clouds, singing. The salesman says, "Looks pretty boring." Then St. Peter shows him the videotape of hell. It is a big toga party, people are feasting, drinking, dancing, flirting. The salesman says, "I'll take this one." He enters through the doors into hell, is hit with a blast of sulfuric air, and is jabbed with a pitchfork. He turns to the devil and says, "This isn't what I signed up for." The devil replies, "You must have seen our demo."

The hype surrounding computer sales and the specially packaged demonstration units showing glorious images of hardware and software applications directly contradict the hardships that users experience when putting computer systems into practice. While computer sales-people can be unscrupulous in making promises, and many a manager can admit to being taken in by them, computers have benefits to offer. But to make them perform in practical work contexts, people must cope with the restructuring problems engendered by the new technology and still get the work done. Where mixed-use and differential-use practices make the adaptations successful, most often the new technology gets the credit, and the dreams of brighter futures through technological innovation are sustained. Why do engineering designers put up with such restructuring and adaptation? The following chapter addresses this by revisiting the role of standardization in engineering history and its relationship to the aura of "high tech."

8

The Aura of "High Tech" in a World of Messy Practice: Standardization, Mystification, and Glamour

You can take a high-tech process and build a piece of crap. It's not high tech because you use the process.

—*An aerospace engineer*

The term *high tech* is used many times in the conversations and interviews discussed in previous chapters.[1] The statement above was made by an engineer who designs sophisticated control systems for jet aircraft. This particular engineer did all his design work with pencil and paper. When I asked him for his definition of *high technology,* he made the explicit distinction quoted above, defining the term as referring to a product and not a process. He further explained that his systems are *high-tech* products because he designs them using *high-tech* components. He maintained that the type of process used to design the control systems—whether pencil and paper or an extensive computer-graphics system—is irrelevant to whether the product or the work that produces it is *high tech* or not. This definition kept the product of his own work within the high-technology designation. In his words:

The hardware we're using is the most currently available devices made by the high-tech companies—Motorola, Texas Instruments, National, all those folks. . . . We don't make the integrated circuits. . . . We take those chips and combine them together as a functional element. . . .

People might associate computer graphics with high technology, where there's no guarantee that they're associated. One, you're talking about what exists and one's a process. And you can certainly use different processes to accomplish the same end. . . . You can take a high-tech process and build a piece of crap. It's not high tech because you use the process.

Why should a control-panel design engineer care about whether his work is classified as high tech? More is going on here than a technical designation. The engineer's self-serving definition of high technology emphasizes product over process so that the outcome of his work and the products of the company that employs him qualify for the high-technology designation. Just as he plays down process to maintain the high-tech definition of his own work, he also ignores the work processes used to create the chips employed in his systems. He categorizes all processes in the black box of the high-technology designation by listing the companies that produce the chips, assuming the listener knows their high-tech reputation. Given the interesting issues raised by his comments, I began collecting engineers' and designers' definitions of *high technology* as I continued my fieldwork. I turn to it here in addressing the next question: Why do design engineers put up with the multiple restructurings engendered by the implementation of computer graphics?

What is happening with computer-aided design is not an isolated issue but is part of both the aura of high tech and the long-term history of the codification of knowledge in engineering. CAD is supposed to codify tacit knowledge and hence to eliminate the need for it. Engineers are not opposed to CAD because the very history of engineering depends on such codification. However, as has been shown in previous chapters, the knowledge that is actually used to get the work done—tacit knowledge, including local, visual, and kinesthetic knowledge—cannot be totally standardized, codified, and ordered to the point that the flexibility of creative work is hampered. Much of the tacit knowledge of design engineering is embodied in the action-centered thinking that takes place using pencil and paper. By its very nature such nonverbal and nonmathematical knowledge can never be totally codified; a residual amount, crucial to design work, will always remain tacit. But the computer-graphics industry maintains an ideal that holds that everything in the design process can be captured and codified. This ideal that perfection is possible is perpetuated by the aura of high tech—the never-ending quest for the constantly elusive, newest, latest, and greatest technology that loses its glamour as soon as it is widely understood. Members of the world of engineering design adhere to this ideal because they feel they ought to believe it even though their experience has shown them otherwise. The aura of high tech constructs a vision of a great future, and because engineers want to subscribe to this source of cultural power, it is difficult for them to criticize it.

Engineers want to use computer graphics; their managers and corporate marketing agents want them to use computer graphics. At present and historically, engineers are not and have not been resistant to new technologies or to the standardization of their practices that machine use entails. The professionalization of engineering has been built on the standardization of parts, processes, and people, and the cultural value of contemporary engineering has been linked to its claims to being high tech. Engineers are not Luddites and do not dislike codified knowledge, which has served them and the profession as a whole so well. They seek out mixed practices to incorporate new technology in design work and are ready to fill in the gaps to make the machine work despite the monumental problems and restructurings that accompany the introduction of the new tool. Just as engineers use mixed practices to protect the familiar cognitive spaces where visual creativity takes place, they also protect engineering by keeping it technically up to date, which is why the aura of high tech is important in perpetuating not only the mystique of engineering but also support for further codification and computerization.

The History of Standardizing as a Contributing Factor

Most approaches to evaluating CAD capabilities and organization do not consider it to be as a long-term process or a process that is part of the codification of knowledge of engineers, central to the occupation itself. This is why looking at the history of engineering is an important analytic turn. By paying attention to history, we can contextualize the implementation of CAD in terms of the historic development of engineering rather than see it as simply the problems of interacting with a new technology. For that reason, I briefly recapitulate it here.

Chapter 2 used both the history of professions in general and that of engineers in particular to show that part of the professionalizing process is codifying, standardizing, ordering, and rationalizing a knowledge base in order to claim it for the market (Larson 1977; Freidson 1986). It was nineteenth-century technical school educators' codification of knowledge from the machine shop floor into formal knowledge that claimed the market monopoly on engineering knowledge for their graduates. At the same time, both new school-trained engineers and shop-trained engineers were interested in the problem of systematizing and standardizing materials, measuring methods, tools, and parts. While both the ASME and technical journals expressed interest in rationalizing the nomenclature of the machine and shop, such rationalization was mainly the result of a generation of textbooks and handbooks (Calvert 1967).

The late nineteenth century witnessed battles over who should set engineering standards and how they should be decided, along with whether the metric system should become the U.S. standard. The early twentieth century saw Taylorism introduced in the attempt to standardize work practice itself from an engineering perspective. The assumption that computer graphics can provide the one best means of getting work done echoes the same mistaken premise of Taylor's scientific management that ignores the messy nature of work and the constant patching up of misunderstandings in communication. Nevertheless, standardization has played a major role in establishing engineering as a legitimate profession. It has not served a single universal good but has served the best interests of various parties at any given time, place, and practice. Indeed, not just the standards themselves but also who has control over setting them is a major variable in who reaps their benefit or who can claim the knowledge so codified. The latter is one point of contention in the implementation of computer graphics. As was described in chapter 6, CAD is now given the credit for the work of drafters, and their jobs have become denigrated, although seasoned veterans still fill in the information gaps that graphic systems and computer operators overlook. Engineers also fill in gaps in the new technology by employing mixed-use practices, incorporating elements from both the paper and electronic worlds. Here, too, though, the technology is more often credited than the resourceful people that make it work despite its shortcomings. Why is this so? This chapter looks at how the glamorous high-tech aura causes people to lose sight of their own as well as others' contributions to making technology work, as illustrated by the comments of the aerospace engineer at the beginning of this chapter and his self-serving and unquestioning reverence for high technology.

The Aura of High Technology

The root of the word *technology* is the combining form *techno*, borrowed from the Greek for "art or craft." The *Oxford English Dictionary* (1989) traces the English usage of *technology* to designate the scientific study of the practical or industrial arts to 1615 and its use to designate the practical arts collectively to 1859. The dichotomous pair *high technology* and *low technology* are discussed in comparison; the earliest usage is listed as 1964.[2] Hence, the history of the term links it to art as skill and to practical arts, certainly not to the high art patronized by churches and kings but rather to the low arts of craftsmen.

In current usage in the everyday world, including industry, the addition of the word *high* suggests that technologies so designated are not unlike other *high* designations in our society, which historically and contemporarily rely on high-low dichotomies for their status. Examples are *high culture, high art, high fashion, high church,* and *high class.* The implicit assumption is that the designated *high* artifact has more status than the unnamed low or middling other. The designation of *high tech,* like the designation *high art,* has definite status overtones that only marginally relate to functionality.

For past general populations and for today's art community, an art work judged by taste makers to be worthy of museum display becomes a symbol of power and prestige for both its owner and creator.[3] In contemporary cultural studies, such high-low or high-popular dichotomies have fallen from favor (Crane 1992), and the relation between social status and taste is highly debated (Zolberg 1992). Researchers have documented that at least in the United States upper-class economic status does not necessarily correlate with a taste for high art (Gans 1985; Halle 1992, 1994; Lamont 1992). This chapter raises the possibility that high tech has taken over from high art the ability to signify power and prestige, at least in industry. Perhaps the technocrat of the rationalized world, who Weber lamented had replaced the "cultivated man," has dealt high art the final blow by elevating taste for and possession of high technology as the status replacement for taste for and possession of high art—at least in her or his bureaucratic domain. Hence, new products and procedures designated as high technology confer status on their corporate possessors and producers not because the new technology may get the job done more efficiently (it may or may not) but because it signifies power and prestige within the pertinent social networks.

An important element of such status for a new technology is its temporality, in that it is crucial to be the first or at least one of the first to possess a high technology. This parallels the status accrued by possession of original art objects. Walter Benjamin (1969) noted that an aura is present in the original work that is absent in all reproductions because the uniqueness of the original derives from being the original for reproductions. Just as mechanical reproductions—engravings, lithographs, and photographs—came to enhance the value of original artworks through dissemination because they represented the work but were not the work, new technologies become known beyond their original context but are of limited availability because of high cost or intentionally limited access. During this period they develop a heightened aura—Benjamin's "presence of the original," which he attributes to acknowledged artistic masterworks. Initially, the making of copies adds to the notoriety of the original new technology, while the time lag involved in dissemination also enhances the aura. The aura then gradually begins to fade as the new technology is demystified through dissemination of the copies. Further widespread dissemination eventually kills the aura as workers work out, around, and through its bugs or inadequacies and incorporate the new technology into daily work patterns where it becomes mundane. Hence, a high technology is elite today and obsolete tomorrow, rapidly losing its status to a newer or higher technology. A technology possesses the aura of the original only when it is new and the problems incurred in its actual application to daily production are not yet solved.

The actual activity of working out the problems of applying the new technology to daily work diminishes a technology's aura because such activity leads to increased understanding and diminished mystification. By the time those using the technology resolve the problems or develop ways to work around them, the product is no longer the so-called latest and greatest of corporate computer jargon but merely part of daily routine. Nevertheless, the glamour of the aura initially overshadows the ponderous task of working out the problems in the workings of the new technology itself or making a fit with preexisting work practices. So, too, its glow blinds those enamored of it to the day-to-day work that goes into designing and producing it. The design work that produced the technology is mystified in that it is seen as the product of geniuses or at least of special persons with access to obscure knowledge. I return to the subjects of mystification and secret knowledge in the discussion of informants' definition categories.

Symbolic Tools

The aura of a high-technology designation delegates a certain kind of status to a tool or technique and hence renders products and processes so named as symbolic tools recognized more for the status they confer on their users, owners, and makers than for their actual function in getting work done. Anthropologists have recognized symbolic tools in non-Western cultures for some time. Symbolic tools bestow power and status on their possessors not because of their functional capability but because of their status as special objects from special sources and special places within the fabric of the social network including its hierarchy, lineage, and history. For example, Thune (1983) reports that for the Duau people, the *palelesalu* (an oversized, greenstone axe blade) is among the most common valuables used in exchange ceremonies and often is a family heirloom. Lepowsky (1983) reports that perhaps the most essential ceremonial valuables on Sudest are greenstone axeblades called *tobo-tobo*, adding that there are many thousands of them in circulation. Before the nineteenth century they were used as tools and weapons, but today they function ceremonially. According to Lepowsky, Sudest people say that five *bwam* (a small axe blade) equals one very long axe blade, just as European money has different denominations. Particular blades called *roguigu* are treasured as magical aids said to be able both to "call" other valuables to join them and literally to give birth to valuable larger blades. While New Guinea tribe members may still use tools such as ceremonial axe blades in a functional manner as well as trade them, the power and prestige the axe blades bestow within the culture comes not from their functional capability but from their position in the social world.

An example of the symbolic use of the high-tech designation can be drawn from contemporary engineering practice. As mentioned, earlier, during the design meeting in which engineers and technicians discussed the choice of a vender to mold a newly designed surgical instrument, one engineer specifically stated, "What's nice about this is we're not using any new technology. We're just putting things together." Later in the meeting the same engineer discussed the virtues of one vendor against another with the rest of the design team. One criterion mentioned was that company A had the ability to do high-tech jobs while companies B and C did not. The design team eventually chose company A despite its higher price. Though the ability to do high-tech production was not the only criterion used, the fact that it was included illustrates the symbolic function of the high-tech designation. Those responsible for making the decision acknowledged they did not need the newest materials, processes, or concepts—in other words, high technology capabilities—for the functional production of their product, yet the technological abilities of the top-rated company became symbolic of that company's technical quality. Similarly, the aircraft control-panel designer quoted earlier justified the high-tech status of the chips he used by referencing the high-tech reputations of the companies who produced them.

Shared Characteristics in Informants' Definitions

While informants tended to define *high technology* as something to do with computers and in overlapping terms of process, product, and concept, more revealing were the characteristics that permeated almost all the definitions—glamour and mystification. Informants almost always included a list of comparative and superlative adjectives, such as *newer, better, faster, smaller, more reliable, more efficient, easier to use, higher quality,* or *highest quality* in their definitions. The constant use of such comparatives in informants' definitions of high technology illustrates a perception of infinite refinement over the present state of any given technology regardless of its current sophistication. This infinite thrust into the future invokes the mystique of the unknown. Indeed, the linking threads between the various definitions of high technology were constant references to the mystique of unattainable, unknown, or hidden knowledge as illustrated in this civil engineer's dual definition:

I've got two definitions of high technology—my own personal and my own public one. My own public one is high technology is any technology you don't have. . . . Believe me, this [computer graphics in drafting] is very much high technology in our organization, even though plants that we design and build are high technology compared to other firms or other, um, or even other engineering disciplines. . . .

My personal feeling is that high technology is things that are new—and complex. . . . High technology is anything I don't understand right now. It's, ah— once I understand it, then to me it becomes routine, and so not, not in that world that's out there that I don't understand anymore. So if I apply that definition to our treatment processes, it would be things that are unique, that we don't do everyday, that have some risks, that have some complexity that we typically don't do.

The civil engineer's definition of high technology as "any technology you don't have" and as "anything I don't understand right now"—coupled with his statement that "once I understand it, then to me it becomes routine . . . not in that world . . . that I don't understand anymore"— illustrates the concept of high technology as a promise of something unknown, secret, mystified—to be delivered in the future—as opposed to the familiar, the routine, the mundane. Hence, high technology can be understood to be similar to John Berger's (1972) description of publicity images: they "belong to the moment in the sense that they must be continually renewed and made up to date." Berger maintains that publicity makes the single proposal "to each of us that we transform ourselves, or our lives by buying something more" (1972, 132). According to Berger, the future promises this transformation through the example of those who have been transformed and who, as a result, are enviable. The state of being envied constitutes glamour. Like advertising images, the concept of high technology never speaks of the present but always speaks of the future. This is the glamour of high technology and the source of its aura. Sexist advertising practices illustrate this at the extreme, promising future sexual fulfillment: "Buy this car, this shampoo, this deodorant, and he or she will fall at your feet." Indeed, glamorous new high technologies are often spoken of as "sexy." The aura of high tech makes a similar kind of promise: "Buy this latest greatest product, and you will become a leader in industry." The promise is based on constantly looking to the future for fulfillment, power, and leadership—all glamorous promises for tomorrow. And tomorrow will bring only more daily mundane tasks along with more promises.

Since the fulfillment of the promise can never be achieved, informants tended to give definitions that could accomplish something in the present. Their definitions portrayed their work and their employer in the best light possible in terms of the status that they perceived was from the high-tech designation. At the end of each of his definitions, the civil engineer quoted above adds a justification for the high-tech quality of his firm's work. In the first statement he does this by contrast: "plants that we design and build are high technology compared to other firms or other . . . engineering disciplines." While he states that high technology is anything that he as an individual does not understand because it is new and complex, he grants a higher level of understanding to the firm in that its new treatment process for sewage is unique, complex, and risky—something "we don't do everyday." The last statement is significant since daily practice kills the aura of high tech. Yet both of his definitions ignore the space between the promise and the delivery of the promise. While many managers have become wary of the hype of computer salesmanship because all too often they have had to purchase more hardware and software to make a product work, the high-technology designation still retains the mystique of the promise of tomorrow and clouds the hard work that is needed to make improvements in any field.

In a similar example, the chief engineer of the computer applications group at a company that builds industrial complexes ranging from nuclear power plants to cookie factories described high technology in terms of the complexities of integrated automation processes—a description that put his company ahead of others despite acknowledged problems. Though he acknowledged that work at his company was less automated than at other firms, he discounted this by reference to the more complex nature of his firm's projects. He based his view of automation and its future on the incorporation of various automated processes into one large, integrated system that would encompass all the aspects of designing a complex industrial installation:

See, I have high expectations, so I don't consider computers around the office high tech, um, I don't consider "walk through" high tech. Um, I think data sharing is at a very, very sophisticated level where all the people are smart enough to know how to use the system that can do that. That would be high tech to me. . . . I've been around so much of this stuff by now that I don't think anything that we have is high tech. I think it's all pretty simple stuff. Integration's probably high tech is my answer. . . . CAD evolved around mechanical. It started as a mechanical thing. And the reason it did is because it's easy to do a widget this big and all the emphasis has been put into mechanical. And in the business we're in, this big broad business, we lag way behind these people because the problem is so complex, and it's going to take so much work to get it done that it's sort of like—. For the people that go to conferences and I listen to mechanical people they just don't realize. They have no idea what real problems are. . . . But man, this field, this is awesome, it's humongous, it is big. And when you think about what we're doin' sometimes. We're building a billion dollar facility. This thing for New Zealand for instance. . . . It was three miles by three miles. . . . The system I can see in my mind today will not be in place when I die. I mean I'm convinced of that 'cause I've seen ten years go by, and boy, have we . . . we have creeped [sic] along. You wouldn't believe how ancient, how slow, this process has been. In my mind it's just incredibly slow.

The chief engineer's expression of "high expectations" for tomorrow simultaneously illustrates the dying spiral of yesterday's high technology, which continually loses its status to become "all pretty simple stuff." And yet this same graphics manager complained long and loudly during the interview and at a previous computer-graphics workshop about the daily problems constantly being worked out in the complex system he oversees. For instance, various customers and vendors wanted their graphics done on different systems and in different formats, resulting in compatibility and scheduling problems for his department. Everybody else's problems are small compared to his. He even acknowledges the great amount of daily work it will take to make graphic systems and integration work in his statement that "the problem is so complex, and it's gonna take so much work to get it done." And yet despite this acknowledgment, the aura, the mystique, the promise of futuristic computer technology prevails. Despite the problems of application today, he concludes, "The system I can see in my mind today will not be in place when I die." The implication is that it will be—some tomorrow.

Mystification

While the promise of tomorrow can be glamorous, so too are the secrets of today—through mystification. A systems designer in a branch of electronic communications that protects satellite signals introduced the role of secret knowledge, alluding to what she called her "secret project" in language that enhanced its mystification:

High tech to me, is, especially in communications, is something that you do particularly fast or better or cleverer than another competitor or something. I don't think of it—I mean, we do use very high tech, quote unquote, latest, greatest type of micro processors and things to do that. . . . I think of high tech as using some really new . . . things to do it. . . . But I don't think of it in terms of anything mechanical or production wise or anything like that. I think of it totally in terms of design concept. . . . I think high tech is if you use a very advanced architecture or something. Isn't that strange? That's what we do.

While her definition of *high technology* puts concept foremost, tying together process and project components, the comparative element is still present in her statement "something that you do particularly fast or better." She appears insightfully to recognize that her definition of high technology is the definition of what she does, perhaps realizing the self-serving nature of her statement that high technology is "us[ing] a very advanced architecture" since she adds a comment: "Isn't that strange? That's what we do." She continues to use such language as she continues:

I kinda tire of tracking the IC state. I'm kinda more into the systems high-tech stuff. (laughter) That's really weird, and actually, . . . the secret thing I do is the protection of this data. While there's more latest, greatest of those, too, that's kinda too high tech, but it's something that most people have never heard of, they're really nasty little devices, (laughter) so people go "big deal." (laughter). Well, if you worked in it, you'd love it, you know.

References to the mystification of high technology show up throughout this communications systems designer's speech in statements such as, "That's strange," "Isn't that strange?" and "That's really weird," which surround her comments about the "secret thing" that she designs.

Her project is secret because of military and commercial security practices. The limited access to knowledge maintains monopoly while adding another dimension to the mystique of high technology. Not only is it mystified simply because it is "anything I don't understand right now," in the civil engineer's words—that is, advanced knowledge that requires training and experience—but here it is knowledge that is intentionally kept secret. Whether the design concept of the product is simple or complex is immaterial. Its mystification is maintained because the basic components of its design are intentionally obscured through limited access—access controlled by the politically or economically powerful.

Conclusion

The aura of high tech and the use of new technologies such as CAD as symbolic tools in engineering continue to add status through the mystification of the mundane and messy work practices that are necessary to accomplish the goals of the job. Additionally, the history of standardizing and codifying campaigns in the development of engineering reveals how intrinsic such activity has been to the professionalization process.

The engineering world must protect its hard-won gains as professionals and embrace the new technology of computer graphics into its domain. CAD is, after all, a system of codification and adds to the high-tech aura of the profession. And the standardization of practices imposed by CAD does not conflict with the growth of the profession since the historical purpose of all this codifying activity has been more to establish the market monopoly and high status of the profession than to actually get the work done. Since engineers are perceived as developing new technologies, their keeping up to date seems inevitably to require the use of CAD. But engineers must also protect the tacit domain of uncodified knowledge if they are to continue to be able to do their work. The use of mixed practices accomplishes both goals.

The aura of the high-tech categorization glamorizes new technology or updated technology, generating a sense of awe not unlike that articulated by the *New York Times* in heralding the "lightning lines" of the telegraph as the "divine boon" and "wondrous event of a wondrous age" (Czitrom 1982). The high-tech glamorous aura is often perpetuated through the use of computer graphics for design presentations. While impressive graphics may be used for sales promotions and illustrations in customer-oriented literature, actual conceptual design work is often done with pencil and paper, and the more painstaking detail work is allotted to computer graphics operators.

The question that began this chapter was, Why should a control-panel design engineer care about whether his work is classified as high tech? It is echoed by informants' definitions that turn out to be either self-serving definitions of what the informants themselves or the companies that employ them do or a mystification. One answer to the first question is that people care that their work and their firm is designated as high tech for the same reason that someone buys a Michael Graves tea kettle instead of a Revere Ware one. Both boil water, and perhaps the Revere Ware kettle does so more efficiently. But the Michael Graves kettle is more interesting to look at and carries status as art in the kitchen. Similarly, the previously mentioned medical instrument design team chose a vendor company that was known for its high-tech work even though the latest and greatest technology was not needed for the project. In these cases, people are relying on accrued status that has been socially constructed by a social network and that takes a recognizable symbolic mode bearing the label of high technology.

Such designated high technology has practical function—just as Michael Graves kettles can boil water and ceremonial axe blades serve a function to the Sudest people of New Guinea. But practical function alone is not sufficient explanation in this case. Something else is at work, which I believe is high technology—functioning as a symbolic tool. Because of the glamorous aura of high technology, we often forget the human agency that goes into computer design work and the human agency that is required to fit the new technology into work settings and practices.

Just as engineers use mixed practices to protect the cognitive spaces where visual creativity takes place, they also protect engineering by keeping it technically up to date. To continue to use pencil and paper in a world of CAD could be seen as analogous to espousing your love of country and western music during intermission at the opera. But just as we live in a multicultural world, where art openings may indeed play rock-a-billy music, to actually live in a high-tech world and get the work done means to practice in a multitech world—to play on all levels. The symbolic importance of high-tech tools makes it easy to overlook experiential knowledge. But there will always be a need for low technology in engineering design because design is tied to tacit knowledge. Designers using CAD systems over time may develop new forms of nonverbal knowledge through interaction with their new tool, though those may be very different from and serve different purposes than the nonverbal forms now being used. We must wait and see. In the meantime, those engaged in design cannot throw away the local knowledge crucial to their work.

Indeed, some of the most creative work being done is not by those who design new computers and software but by those who use them creatively, preserving older paper-world practices, where appropriate, and applying computer technologies to repetitive tasks. We need to recognize the creative resilience of design workers—engineers, designers, drafters, and their management—in their innovative mixed-use practices that fill in the gaps in the world of electronic designing and drafting. To do so challenges the glamour and the aura of the new technology. Being aware of the messy daily practices of sketching, drawing, and turning back to paper copies to red-line them—in contrast to crediting the dark screen with the sparkling colored lines with accomplishing the task—is to elucidate the mystery and reveal the hard and messy work of getting the job done.

9

The Power of Visual Representation: Mixed Practices, Multivisual Competencies, and Meta-Indexicals

Why are visual representations so powerful? I have suggested that it is their *meta-indexical* quality—their ability to be a holding ground and negotiation space for both explicit and yet-to-be-made-explicit knowledge—that allows them to be more than the sum of their parts as well as more than Latour's "center of calculation." Latour (1986) makes the point that it is not the inscriptions that are interesting but rather their transformation from the empirical to the theoretical, from the informal to the formal, as more and more information accumulates.

The data presented in this study show that one very important capacity of the visual lies in its malleability—its ability to be drawn interactively and shaped and redrawn and reshaped by members of an engineering design group. In this process, the visual representation integrates and informs the collective and changing cognition of those designing it. Equally important is the particular way visual representations facilitate the joining of not only multiple meanings but multiple forms and formats of coded and uncoded, verbal, visual, mathematical, and tacit knowledge. This ability to serve as a gathering ground for multiple ways of knowing is the meta-indexical property. Visual meta-indexicals have at least the ability to:

• Transform other ways of knowing, such as verbal and mathematical modes, into a visual format;

• Index, reframe, or simply reach the unarticulated or tacit knowledge of interacting participants;

• Elicit tacit knowledge from participants so that it can be represented in a format readable to others;

• Represent knowledge through the flexibility of sketching in an uncoded format;

• Tap all sorts of visual modes of representation spanning all of the world's art forms and history, including the abstract (what I have referred to as *multi-visual competencies*);

• Expand a given visual lexicon so that new codes can be incorporated;

• Develop and standardize a new lexicon to maintain consistency of meaning (Latour's immutability); and

• Represent different ways of knowing using many different systems of representation at once, including verbal, mathematical, and numerous visual modes.

The joining of these and perhaps other qualities allows visual representations to shape the structure of design work while controlling who participates in the design process.

It is because of their meta-indexical qualities that drawings and prototypes can so easily function both as boundary objects that are capable of being read on different levels by different groups involved with the design and its final product as well as conscription devices that are so central to the design process that they must be used by anyone who communicates about the process. The role of visual representations in facilitating the creation of designed products can be more clearly delineated when they are understood as meta-indexicals, boundary objects, and conscription devices. These concepts help elucidate how visual representations facilitate group thinking and consensus among design participants and also help explain why they sometimes become the turf of contention when conflict occurs in the design-to-production process.

Bucciarelli (1994) also notes that design is a process of achieving consensus among participants who have different interests and who bring different perspectives colored by different histories to their work. Participants must negotiate these differences, bringing coherence to their perspectives and fixing them in the artifact. Bucciarelli concludes that there is no overriding perspective, method, science, or technique that can control or manage the process through which participants negotiate their differences. Visual representations do not necessarily provide such a structure, but they are an arena in which negotiations can be held. They allow intangible ideas to become concrete—but still allow ideas to be reworked and renegotiated. As boundary objects, each new concrete piece has shared meanings as well as individual meanings for the participants. As each concrete piece becomes represented and negotiated, it invokes further pieces for negotiation, with the visual representation providing the centering device around and through which the negotiation takes place.

This use of visual representations parallels what Suchman (1988) describes in her analysis of the group use of the whiteboard by cognitive scientists, but it is still more powerful because it indexes knowledge beyond what is actually represented. Suchman characterizes the whiteboard as structuring a "shared interactional space" (319) and a "second interactional floor which is co-extensive and sequentially interlocked with talk" (322). Engineers' drawings, like the scientists' white boards, "achieve a conceptual ordering between items" (320), although the ordering in engineering drawings is often a spatial one. Both the drawings and the whiteboards may be delineated into owned territories or inhabited jointly. The merging of individually owned and interpreted territories, as well as jointly inhabited ones, into a larger jointly inhabited frame is the very function of representations as boundary objects. This merging allows engineers' visual representations, like the scientists' whiteboards, the flexibility to become an arena for the introduction, manipulation, and resolution of design dilemmas.

Visual representations and linguistic representations are not the same, but comparisons are useful in teasing out just why visual representations are so powerful—not to explore an analogy between language and visuals but to consider the culture-based acquisition of literacy in a lexicon-based code system. Art historians suggest that a culture comes to see the world in a way that is linked to its material experience of rendering it (Alpers 1983; Baxandall 1972). The historical development of drafting conventions and the examples of resistance to change in those conventions presented in this book indicate that engineers and drafters do indeed have a visual culture. Within that culture they have developed multiple literacies in reading a great variety of formal and informal visual lexicons, many of which I have discussed and illustrated. All of this points to visual literacy as interactively learned knowledge that is situated in practice, just as Lave (1988) and Suchman (1987) have suggested for other types of knowledge and Zuboff (1988) has acknowledged as being present in work in general. Visual literacy is not a monolithic universal but is rather based on culture and experience. Visual cultures and visual literacies are made of infinite sets of different lexicons. These are sometimes layered or intermingled with one another within a boundary object or conscription device, allowing knowers of different types of knowledge to coordinate work and conscript interaction using the visual representation as a vehicle.

The collective cognition or collective visual knowledge I am discussing here is certainly a social construct. It is interactively negotiated outside the confines of the individual, where a sociologist can witness it. At the same time, even when it is not represented on paper, our own experience tells us that some manipulation of the visual dimension can take place internally in our mind's eye, as Gardner and others suggest. Lave's work on mathematical knowledge as situated in practice can be viewed as a link between the public negotiation of knowledge and some internalized, individual negotiation akin to G. H. Mead's (1934) generalized "other." While one could align the perspectives of Latour and Lynch against those of Lave or Gardner as mutually exclusive, they do not need to be. Recall the story of the man who had been blind since early childhood and then had his sight restored as an adult (Sacks 1993). Although he was physically able to see after his surgery, he could not make sense of what he saw because he lacked socialization to his own visual culture. The situated knowledge that ordinarily would make visual clues relevant is of the same nature as the situated knowledge that allows Lave's "just plain folk" (1988, 4) to make computations accurately in the markets of the third world or the grocery stores of the first world without using school-learned mathematics.

When we unpack the residual category of tacit knowledge, we find that some of the pieces of engineering design are visual in nature. We can witness this in sketches, drawings, and the interactions around them. But visual cognition should not be reified into dichotomous categories of the social or the cognitive because the two are interactive. They mutually construct one another in a dynamic, negotiated visual culture that is grounded in practice. If Pasteur's bugs are permitted to be actants in the social world (Latour 1988b), then "the mind's eye" (Ferguson 1977, 1992) can also be an actant that shapes and is shaped by human interaction.

To fully understand this quality of visual representations as meta-indexicals in situated context, we must briefly retread the ground we have covered. We have explored design engineers' historical precedents and their contemporary practice in light of their visual culture and their standardization heritage. We have looked at their resourceful, mixed-use practices that facilitate the advantages of both the paper world and the electronic world. We have seen how such mixed practices help designers deal with the proliferating restructurings that computer graphics introduces into the work, workgroups, and company organization of design engineering.

Computer graphics are an excellent tool for manipulating codified knowledge, but many designers find hand sketching quicker for noting down ideas and find paper copies more accessible for analysis. Hand sketching is an excellent tool for manipulating uncodified knowledge but also can incorporate forms of codified knowledge. Drawings using drafting conventions employ codified and uncodified knowledge. The key is that when we consider the humans interacting around explicit, visually represented knowledge, other multiple forms and layers of tacit and local knowledge can also be elicited, captured, and represented to some degree.

Sketches are the real heart of visual communication. The case studies and the comparative data presented in this book have shown their importance as carriers of visual knowledge because they serve both as individual thinking tools and as interactive communication tools. Recall the drafter who complained that she couldn't think without her drawing board. Sketches allow individual thinking and interactive communication to occur simultaneously and hence facilitate distributed cognition. Thorough designers continually use sketches, ranging from early drafts that are discussed with fellow designers and fabricators to rough drawings that are sketched in margins to clarify an idea.

The nature of sketches allows them to expand beyond a given lexicon, grammar, or syntax. Not only can sketches incorporate lexicons and grammars from many engineering specialties, from specific industries or companies, and from the entire cultures of fine, folk, popular, and commercial renditions. They also can be used to invent and employ newly invented symbols and rules to govern them. It is as if in verbal communication one could speak or write using the vocabulary, grammar, and syntax of standard English; work in some French and Russian phrases as well as some newly coined words, phrases, and grammatical structures; and then expect monolingual English speakers to follow the meaning.

This is not to say that sketches alone convey meaning. Their use is embedded in context and language. But so also is language dependent on visual representation and context, all of which are interlinked with tacit knowledge. Joined with language and tacit knowledge, sketches achieve a universality broader even than that of a language code. One can think of it as a meta-universal or panlexical code, which is why I have chosen to call it *meta-indexical.* This extreme flexibility is one of the aspects that makes visual representations significantly different from other forms of communication. It is also one of the reasons sketches work so well as boundary objects and conscription devices in facilitating distributed cognition: their extreme flexibility facilitates multiple readings of the same image by different parties. It is this meta-indexical flexibility that allows creativity to flourish because it both supersedes the universal and restricted codes of drafting conventions while simultaneously employing those standardized lexicons along with other visual conventions, preexisting or invented on the spot.

The visual culture of engineering is more than people turning to drawings to answer a design question or collaborative visual thinking. Other forms of knowledge and communication—verbal, mathematical, tacit—are built around these representations. Since the visual culture of engineers is constructed by all these component parts, in consort, a technological change in the everyday practices of rendering, such as the introduction of a computerized graphics system, will necessarily create profound changes in the visual culture.

The existing visual culture of engineering has been constructed through mental and physical action that places pencil on paper and through drafting conventions built up over time. It is not just that those who work in design engineering have a preexisting visual culture that they have inherited through the traditions of drafting protocols. Thinking is not purely a mental activity but is a physical one as well. What goes on in the mind is tied to material existence and material practice. Engineers, designers, and drafters participate in tangible actions that construct their visual culture, their way of seeing, literally and figuratively. Years of drawing and thinking and thinking and drawing mean that the two become wholly intertwined. It is well known in the engineering world that designers can visualize a whole machine, turning it to examine it from all directions. This is a skill developed through practice that takes place in the situated context of years of drawing. The hype that extols the capabilities of graphics systems maintains that a similar animated view of a new design can now be done on the computer. But this is not exactly the case.

The connection between the designer's mental vision and the one on paper is not the same as the connection between the individual's mental vision and the one generated on the computer. The mental vision and the paper representation interactively construct one another. The process is not whole vision and then, plunk, drawing. The action orientation of drawing—scratching down visual concepts that may occur in fragments and then putting them together as a whole, perhaps after trying various combinations—are part of the thinking process. When design work is done in groups, the interaction becomes collective as two or more people sketch to one another as a form of visual conversation, generating tangible representations of their mental concepts, giving first approximate shapes to ideas and then together combining and modifying them so that both the collective drawing and their own mental images of the design are modified.

In good design work this interactive sketching communication continues even during the final production phases of a new machine. Nevertheless, when I stepped onto the shop floor, engineers who were designing a new turbine engine package were embarrassed that I was seeing the messiness of the process, the sketches in the margins, the further corrections to the drawings as the machine got built and sometimes dismantled and rebuilt. They were embarrassed because I was seeing the contrast between actual practice and the myth of technological development, which is supposed to occur in a smooth outgrowth from clear scientific formula applied to human needs. The same mythology contributes to the hype used to promote the use of computer graphics and its underlying philosophy that there is one best way to do engineering design. But industries, companies, design teams, and individuals differ. While there may be good designs and poor ones, there is no one best way to do design work. Trying to impose one through electronic mediation destroys much of the creativity that design work is about. Institutional variability means that a tool designed as the one and only best method has serious shortcomings. The organization may tell designers not to use the old tools—shortcut methods from the paper world—but resilient designers and drafters use them anyway, in conjunction with the new tool, generating creative mixed practices to get the work done on time and to hold on to some of the familiar cognitive space of a visual culture in flux. So strong are these mixed-use practices that they even spill over into the choices between hardware and software in the differential practices of designers.

Some design activity can indeed be facilitated using computer-assisted design. That identical components, especially in chip design, can easily be duplicated and moved around using a graphics system is certainly an advantage that reduces repetitive work. However, for conceptual work, which is the root of design, a good number of chip designers, mechanical engineering designers, and civil engineering designers report that they usually start with paper. The computer dictates how people must enter information, and designers regard pencil and paper as more direct in getting a fleeting idea down as quickly as possible. Again, this points to the physicality of the link to the knowledge. After fleeting ideas are captured in hand sketches, they can then be entered into the computer, and electronic graphics capabilities can be used to further refine them. Sketches can also be printed out again on paper when designers need to analyze the design as a whole. Again, at such an analysis phase, thinking is connected with the physicality of seeing and manually manipulating visual ideas. People employ such mixed practices to take advantage of the capabilities of the new tool while still maintaining important paper-world practices that are tied to thinking and analysis in visual terms. Use of the computer-graphics tool is just another of designers' multivisual competencies—a very strong one that will affect their practice and hence their visual culture.

Recall the example of the computer-graphics-trained engineering students who collectively continue to make the error of omitting the crucial line over the opening of a cylinder because they do not see construction lines as part of computer-aided practice. Such an error does not reflect a lack of intelligence or skill but is actually the result of differences in practice. The step-by-step practice of electronic-world drafting is different from the step-by-step practice of paper-world drafting. Since those students were encountering a design visual culture in which some of the processes had changed, they were attempting—unknowingly and here unsuccessfully—to change the visual culture. The generation of designers being trained on graphics software is developing a new visual culture tied to computer-graphics practice. It will influence the way they see, and it will be different from the visual culture of the paper world as it combines with that culture to create a new hybrid.

Design engineers embrace computer graphics as a useful tool on some levels and as not useful on others. Standardization and codification of technical information has historically been an agenda of engineering, one that has helped it gain professional status. The high-technology designation, which can be applied to the computer-graphics tool, in many ways is comparable to the symbolic tools discussed in the anthropological literature. The value of symbolic tools is less in their functional capabilities than in the status that accrues to those who use them. The use of computer graphics contributes to the high-tech status of engineering and is in keeping with the standardization and codification traditions of the profession. However, the implementation of computer graphics has spawned geometrically progressive restructurings from the level of the individual working with the design through relationships between groups within companies to relationships between companies who do business with one another. Resistance is to the restructuring more than to the technology itself. The mixed practices described in this book allow designers to embrace the new technology selectively while protecting their tacit knowledge and cognitive space. How the division of labor between paper and electronic practices is meted out varies from company to company because not all companies are alike. Creative people use the new tool to meet their needs and the needs of the company by employing mixed practices that work rather than letting marketing promotions dictate how the tool should be used. The smart managers who let designers do so should witness a new visual culture that capitalizes on mixed-use practices, multivisual competencies and the meta-indexical quality of visual representations that resourceful users are sharing with one another.

Notes

Chapter 1

1. See, for example, Barnes (1982), Barnes and Edge (1992), Bijker (1995), Bijker, Hughes, and Pinch (1987), Bijker and Law (1992), Clarke and Fujimura (1992), Hacker (1990), Latour (1987, 1988a, 1988b), Law (1987a, 1987b), Law and Callon (1988), MacKenzie (1990), MacKenzie and Wajcman (1985), Traweek (1988), Winner (1980) and, in history of technology, Layton (1974), Ferguson (1977, 1992), and Vincenti (1990).

2. Studies of representational practice in science mark the diversity of such visual and verbal formats, which range from graphs, diagrams, illustrations, and models to computer construction programs, any combination of which may be employed to achieve succinct and persuasive presentation of scientific phenomena. See Lynch and Woolgar (1988b) for a variety of discussions on the use of representational devices and their relation to general issues in the sociology and philosophy of science, Law and Lynch (1988) for an exemplary study, Ruse and Taylor (1991) for discussion of visual representation in biology, and Fyfe and Law (1988) for a collection of essays on the more general theme that power and knowledge as social processes can involve technologies of depiction that contribute to the reproduction of the social order.

3. All informants gave informed consent for this research.

4. See Braverman (1974), Cooley (1987), Noble (1977, 1984), and Shaiken (1985).

5. Harper and Sellen (n.d.) discuss crucial at-a-glance documents used in a variety of collaborative work settings. See also Harper, Hughes, and Shapiro (1991); Luff, Heath, and Greatbatch (1992); and Shapiro, Hughes, Randall, and Harper (1991).

Chapter 2

1. Yes, they were probably all and only men by this point, even though previously in early modern Europe women were allowed to be crafts guild members, often at the death of a guild-member spouse.

2. Zilsel (1942) suggested that scientific principles were founded in the same manner, through a codification of craft knowledge. Of course, scientific principles were shaped both by the development of a rhetoric of science that contrasted with the supposedly feminine style of alchemy, as Fox Keller (1985) has shown, and by the evolution of a documentary style for virtual witnessing, as Shapin (1984) has described.

3. Vincenti (1990) points out that practical considerations based on empirical experience can become well codified over time so that they can become quantifiable data. For example, "The limits on sheet thickness for different types of flush riveting started out as rough design rules, but as production experience increased, they became precise enough for tabulation and inclusion in process specifications" (219).

4. Layton (1986) notes that even under Herbert Hoover's less than radical leadership of the Federated American Engineering Societies in the early 1920s, the reports of his scientific management committees on waste and on the advantages of the eight-hour versus the twelve-hour workday were rejected by the business-controlled engineering societies. Ultimately, the efficiency devices of the movement came to be applied and enforced by business management, thus putting Taylorism into the management tradition of organization theory. Thus the revolt of the progressive engineers ended without the profession gaining autonomy from business interests, a continuing aspect of weakness for engineering as a profession.

5. More recent scholarship on the history of American engineers takes issue with Noble's (1977) view of the marriage of engineers and corporate business. Noble is seen as overstating the inevitability and completeness of engineers' accommodations to business, so that he is unable to explain the periodic flare-up of engineering dissent to which Layton and others point. Meiksins (1988) argues for focusing on the complex relations among the diverse forms of engineers' protests over the past six decades while keeping in mind that engineering has been concentrated in organizations for almost a century. He points to more recent conflicts over engineering ethics in the 1960s and 1970s, including whistle blowing, as evidence that engineers have not completely abdicated ethical autonomy to the organizations that employ them. He sees the truly significant result of the 1920s revolt of the engineers as the emergence of a new kind of engineering protest, centered on the material concerns of the average engineer and rooted in the changing conditions of American engineering. Zussman (1985) shows that the identification of middle-class engineers is more complex than that of engineers who have moved into management; the offer of high rewards does not always bring unconditional loyalty, but rank-and-file engineers accept

their lot on more conditional grounds. Whalley (1991) argues that the rapid growth in the need for technical workers has worked against proletarianization.

6. This conflict has even regenerated talk of unions among engineers to prevent deskilling; see Shaiken (1985).

7. The growth of the science-based electrical and chemical industries in the late 1890s heightened the demand for standards of measurement in materials. Noble (1977) notes that MIT president Henry S. Pritchett, the astronomer responsible for developing the technique for establishing standard time, maintained that a centralized standardizing mechanism was vital to the nation's defense against industrial competition from Germany. Pritchett acknowledged that the strongest government support for a new bureau of standards came from influential banker Lyman Gage, whose rhetoric ascribed to national standardization the might of legal and moral authority backed up by the legitimacy of scientific truth.

From his position as director of the Coast and Geodetic Survey in 1897, Pritchett, together with prominent physicist Samuel Stratton (later also an MIT president), Assistant Secretary of the Treasury Frank Vanderlip (vice president of National City Bank of New York in 1901), and Secretary of the Treasury Gage encouraged Congressman James H. Southard to sponsor their bill for the enlargement of the office of Weights and Measures into a more efficient Bureau of Standards. Strongly endorsed by professional science associations, the bill was passed overwhelmingly and the bureau was established in 1902 (Noble 1977).

8. Freidson (1986, 201–202) quotes a Federal Trade Commission report that provides a glimpse of product standards and the process by which they are created—a phenomenon of staggering scope and complexity.

9. The extent of industrial influence on such standards can be seen in the contrast between the standardization of electrical engineering and chemical engineering (see Noble 1977). In the electrical industry, dominated by a few large corporations, standardization went smoothly as staff at General Electric and Westinghouse worked with the American Institute of Electrical Engineers standards committee to develop standards and equipment specifications that would eventually be adopted nationally and internationally. In contrast, the fledgling American Institute of Industrial Chemical Engineers committees on standardization of definitions and specifications progressed slowly because of a lack of interest on the part of business. Noble points out that manufacturers practically ignored their requests for data. By the time the United States began to gear up its industrial resources for World War I, a vigorous national standardization movement was flourishing as the various engineering societies under the American Engineering Standards Committee were joined by the U.S. Departments of Commerce, War, and the Navy, lending the organization a quasi-governmental status.

10. Corporate industry established in-house training programs, coordinating their activities through the National Association of Corporation Schools (NACS). General Electric led the way with its TEST Course, by which the company instilled "the seriousness of business" in college graduates, giving them the "proper conception" of business values. Because of the growing importance

of the electrical industry and its need for college graduates, the influence of electrical engineers in the engineering profession grew, and engineering educators also saw an increase in status as an electrical-engineering curriculum distinct from that of mechanical engineering became necessary and representatives from the industry actively participated in the curriculum discussions (Noble 1977).

11. Throughout the summer of 1918 Mann lobbied for a plan known as the Student Army Training Corps (SATC), which would allow students between the ages of eighteen and twenty-one (draft age was twenty-one) to gain military status by encouraging them to enlist in the program in their school and putting them under the military authority of the CEST. The SATC program was at first voluntary, but CEST lobbied to make it compulsory when the draft age was lowered to eighteen (Noble 1977).

12. Strangely enough, professors did not object to campus militarization. With the authority of the army behind them, CEST leaders certainly displayed impressive power, much more than the academics could counter. Noble also suggests that professors who harbored objections may have kept them quiet for fear of appearing unpatriotic or of being exposed as inferior to the forceful engineers in either mental or practical ability (Noble 1977, 219).

13. The EPCD focused on the engineering student, the engineering college, and the professional practicing engineer. Its long-range goals included a program for implementing the Wickenden report, sponsored by former leaders of the CEST and SATC. Wikenden was supervisor of the GE Cooperative Course at MIT, and his report advocated closer cooperation between industry and education through a curriculum of scientific training accompanied by social science training geared for future management responsibility. The ECPD soon became recognized as the central agency for all matters relating to the engineering profession, including college accreditation, professional standards of ethics, determination of competence for practice, professional recruitment from high schools, and the general definition of what it means to be a professional engineer in the United States (Noble 1977).

Chapter 3

1. A shorter version of this material appeared in *Cultures of Computing* (Star 1995). Used with permission.

2. Arnheim (1969) discussed visual thought as a cognitive activity over twenty years ago. Ferguson (1977, 1992) points out the importance of visual thought throughout engineering history to the present. He notes that Francesco di Giorgio Martini's *Architecttura* (1475) has no meaning without its abundant drawings and that at least one technical codex had no text because words were thought to be unnecessary to elucidate the drawings. Sacks (1993) gives an interesting account of how a fifty-year-old blind man had his sight restored after forty-five years and then had to learn to make sense of the visual world, pointing to the learned nature of visual culture. Studies by engineers also note the role played by the visual: see Bucciarelli (1994) or Vincenti (1990), for example.

3. See Jay (1988a, 1988b) for a discussion of the ocularcentrism of Western culture that embodies its own ideologies.

4. Zuboff (1988, 187) in her study of automation in industry, notes that in various factories "action-centered skill, like all purposeful human activity, requires the intelligent participation of the human brain, but it is an intelligence that tends to be blended with the body's responsiveness and capacity to act." Hence, the development of such skills along with their execution and memory remain confined in the sphere of tacit knowledge. She cites earlier studies that illustrate that few machine operators used fully rational or conceptual approaches and that better operators made decisions based on intuitive understanding based on experience alone. Hindle (1981) documents the necessity of both visual and fingertip, or kinesthetic, knowledge in the transfer of technologies from Europe to colonial America. Sudnow (1978) points out the physical fingertip memory of accomplished pianists.

5. Quoted in Vincenti (1990, 4), which he credits to North (1923).

6. The dependence of formal statements on embodied knowledge was also noted by Fleck (1935). See Cambrosio and Keating (1988) for an excellent overview of contemporary definitions and discussions of tacit knowledge.

7. The informant uses the term *orthographic,* which is synonymous with *orthogonal,* and is the accepted terminology in his company.

8. Suchman (1988) has discussed the whiteboard in other design contexts as a locus of cognitive science practice in that the board both supports and is organized by the structure of face-to-face interaction. By identifying systematic practices of the whiteboard in use, she hopes to understand how those practices and the inscriptions they produce constitute resources for particular occasions of technical work. Bucciarelli (1994, 165–172) too documents that in discussions on design particulars participants turn to the whiteboard and sketch. The final group resolution on the voltage to be used for a residential photovoltaic module occurs when one of the members starts sketching and others collaborate.

9. Vincenti (1990, 207–224) lists the following six categories of design knowledge: (1) *fundamental design concepts,* which are the central operational principles at work in normal configurations for static structures (bridges) or dynamic machines (aircraft); (2) *criteria and specifications,* which are those embodied in engineering standards, such as loads and dimensional tolerances on flushes for rivets or speed, altitude, and horsepower for performance; (3) *theoretical tools,* which are the intellectual concepts as well as mathematical methods and theories for making design calculations; (4) *quantitative data,* which include physical constants, properties of materials and safety factors; (5) *practical considerations,* such as knowledge received from accidents; and (6) *design instumentalities,* including structured procedures and optimization strategies. In his discussion he makes the distinction between mathematics on the one side and more empirically derived data, including a cluster of device-related theoretical tools such as beam theory; phenomenological theories such as the blade-element theory of

propellers; and quantitative assumptions such as how rivets share loads. He notes that such information may be codifiable but is not sufficient by itself. Rather, to be relevant and usable such knowledge "also needs an array of less sharply defined considerations derived from experience in practice" (217).

Chapter 4

1. A shorter version of this material appeared as (Henderson, 1991a), "Flexible Sketches and Inflexible Data Bases: Visual Communication, Conscription Devices, and Boundary Objects in Design Engineering," *Science, Technology, and Human Values,* Vol. 16: 448–473. Used with permission.

2. There are no one-to-one correlations between the mental world and the world of practice; the two are intertwined and interactive. Concepts of mental models, mechanical representations, and heuristics as discussed by Gorman and Carlson (1990), are applicable but are too broad to serve the purposes intended here. Visual practices, including sketches and drawings, that stand between mental models and mechanical representations could be classified as heuristics, but that does not tell us much. Gooding (1990, 181) emphasizes the importance of situated practices but misses its application to visual representations in his suggestion that "successful communication of novel findings . . . depends on finding the right pictorial or verbal imagery." Rather, what takes place in practice is the working out of imagery in simultaneous relationship to the working out of the concepts involved. Gooding points out that the narrative accounts of invention that have come down to us leave out the uncertainties and contingencies that occurred in context. So, too, do the pictorial representations that have come down to us leave out many of the discarded earlier and interactive first approximations, thinking drawings, and rough drafts.

3. In this industry, drafters could work up to the designation of senior designer and from there be promoted to the title of design engineer on the basis of experience rather than schooling. This is becoming more and more rare as the norm today, especially in Silicon Valley, where such a promotion route requires a college degree.

4. Mukerji (1989) points out the low status of technicians and technological innovators in oceanography despite their crucial contribution to the research process. She further notes that scientists with reputations as good builders of research machinery have trouble obtaining funding for more theoretical work. Collins (1974) notes the intentional withholding of tacit knowledge necessary for building Transversely Excited Atmospheric Pressure CO_2 (TEA) lasers when they were a new technology. Cambrosio and Keating (1988) have found that scientists using hybridoma technology in biotechnology research acknowledge tacit knowledge as art or magic, thus mystifying it.

5. The importance of play in learning to use the computer is exemplified in classroom studies, where those who use social networks and play are more successful in learning computer skills than those who read manuals (see, for example, Riel 1989).

Chapter 5

1. This material appeared in a slightly different form in Henderson (1995a), "The Political Career of a Prototype: Visual Representations in Design Engineering Consensus and Conflict," *Social Problems,* Vol. 42, No. 2: 274–299. Used with permission.

2. An example of the latter would be alterations in mold techniques having to do with pressure and temperature during the molding process, which must be recorded in comments marked on drawings.

3. The "fake eyes" referenced here were used only for judging size in the early stages of the design work. Extensive laboratory trials with rabbit eyes, referred to as "model eyes," were undertaken when the design was more developed.

4. The designation of set 1 and set 2 is arbitrary and does not indicate an absence of numerous sketches, corrections, and redrafting associated with each set.

5. Bucciarelli (1994, 159–161) makes a similar point regarding design decisions in that participants' interests shape their proposals, explanations, and understandings and that these are not necessarily the same for all involved. Such concerns derive from individual technical expertise, experience, and responsibilities.

Chapter 6

1. See Allwood and Kalén (1994) for a review of the international literature on the useability of computer-assisted design.

Chapter 8

1. A portion of the data presented in this chapter appears in a different format in Henderson 1998, "The Aura of High Technology in a World of Messy Practice," *The Sociological Quarterly* 39: 4.

2. The earliest usage of the term *high technology* is thought to be S. Miller (Horowitz 1964, 292): "The youthful poor possess limited or outmoded skills and inadequate credentials in a high-technology, certificate-demanding economy."

3. Zolberg (1984) traces the aura of the fine art museum to royal settings, which were intended to testify to their owner's legitimacy and taste. Becker (1982) sees fine art museums as defining the culture of outgroups as parochial, insignificant, or ignoble. Bourdieu (1984) has suggested that those who appreciate work in the domain outside high art are defined by their own taste definitions as inferior.

References

Ackrich, Madeleine. 1992. "The De-Scription of Technical Artifacts." In *Shaping Technology/Building Society*, edited by Wiebe Bijker and John Law. Cambridge: MIT Press.

Ackrich, Madeleine, and Bruno Latour. 1992. "A Summary of a Convenient Vocabulary for the Semiotics of Humans and Nonhuman Assemblies." In *Shaping Technology/Building Society*, edited by Wiebe Bijker and John Law. Cambridge: MIT Press.

Allwood, Carl M., and Thomas Kalén. 1994. "Usability in CAD: A Psychological Perspective," *International Journal of Human Factors in Manufacturing* 4: 145–165.

Alpers, Svetlana. 1983. *The Art of Describing: Dutch Art in the Seventeenth Century*. Chicago: University of Chicago Press.

Amann, Klaus, and Karin Knorr-Cetina. 1988. "The Fixation of Visual Evidence." *Human Studies* 11: 133–169.

Arnheim, Rudolf. 1969. *Visual Thinking*. Berkeley: University of California Press.

Arnold, Erik. 1984. *Computer-Aided Design in Europe*. Sussex: Sussex European Research Center.

Barnes, Barry. 1977. *Interests and the Growth of Knowledge*. London: Routledge and Kegan Paul.

Barnes, Barry. 1982. "The Science-Technology Relationship: A Model and a Query." *Social Studies of Science* 12: 166–172.

Barnes, Barry, and David Bloor. 1982. "Relativism, Rationalism and the Sociology of Knowledge." In *Rationality and Relativism*, edited by M. Hollis and S. Lukes. Cambridge: MIT Press.

Barnes, Barry, and David Edge, eds. 1982. *Science in Context*. Cambridge: MIT Press.

Baxandall, Michael. 1972. *Painting and Experience in Fifteenth Century Italy*. Oxford: Oxford University Press.

Baynes, Ken, and Francis Pugh. 1981. *The Art of the Engineer.* Guilford, Sussex: Lutterworth Press.

Becker, Howard. 1982. *Art Worlds.* Berkeley: University of California Press.

Benjamin, Walter. 1969. "The Work of Art in the Age of Mechanical Reproduction." In *Illuminations,* edited by Hannah Arendt. New York: Schocken.

Berger, John. 1972. *Ways of Seeing.* London: Penguin Books.

Bernstein, Basil. 1971. *Class, Codes and Control.* London: Routledge and Kegan Paul.

Bijker, Wiebe. 1995. *Of Bicycles, Bakelites, and Bulbs.* Cambridge: MIT Press.

Bijker, Wiebe, Thomas Hughes, and Trevor Pinch, eds. 1987. *The Social Construction of Technological Systems.* Cambridge: MIT Press.

Bijker, Wiebe, and John Law, eds. 1992. *Shaping Technology/Building Society.* Cambridge: MIT Press.

Bly, Sara. 1988. "A Use of Drawing Surfaces in Different Collaborative Settings." Paper presented at the Second Conference on Computer-Supported Cooperative Work, Portland, Oregon, September.

Booker, Peter. 1963. *A History of Engineering Drawing.* London: Chatto & Windus.

Bourdieu, Pierre. 1984. *Distinction: A Social Critique of the Judgment of Taste.* Cambridge: Harvard University Press.

Bourdieu, Pierre. 1985. "The Social Space and the Genesis of Groups." *Social Science Information* 24(2): 195–200.

Bowker, Geof. 1987. "A Well-Ordered Reality: Aspects of the Development of Schlulmberger, 1920–1939." *Social Studies of Science* 17: 611–655.

Bowker, Geof. 1988. "Pictures from the Subsoil, 1939." In *Picturing Power, Visual Depictions and Social Relations,* edited by Gordon Fyfe and John Law. London: Routledge.

Bowker, Geoffrey. 1994. *Science on the Run: Information Management and Industrial Geophysics at Schlumberger, 1920–1940.* Cambridge: MIT Press.

Braverman, Harry. 1974. *Labor and Monopoly Capital: The Degradation of Work in the Twentieth Century.* New York: Monthly Review Press.

Bucciarelli, Louis L. 1994. *Designing Engineers.* Cambridge: MIT Press.

Callon, Michel. 1986. "The Sociology of an Actor-Network: The Case of the Electric Vehicle." In *Mapping the Dynamics of Science and Technology: Sociology of Science in the Real World,* edited by Michel Callon, John Law, and Arie Rip. Basingstoke, UK: Macmillan.

Calvert, Monte. 1967. *The Mechanical Engineer in America, 1830–1910.* Baltimore: Johns Hopkins University Press.

Cambrosio, Alberto, and Peter Keating. 1988. " 'Going Monoclonal': Art, Science, and Magic in the Day-to-Day Use of Hybridoma Technology." *Social Problems* 35(3): 244–260.

Clarke, Adele and Joan Fujimura, eds. 1992. *The Right Tools for the Job in Twentieth Century Life Sciences: Materials, Techniques, Instruments, Models, and Work Organization.* Princeton: Princeton University Press.

Collins, Harry M. 1974. "The TEA Set: Tacit Knowledge and Scientific Networks." *Science Studies* 4: 165–186.

Collins, Harry M. 1985. *Changing Order: Replication and Induction in Scientific Practice.* Beverly Hills: Sage.

Collins, Harry M., and Steven Yearley. 1992. "Epistemological Chicken." In *Science as Practice and Culture,* edited by Andrew Pickering. Chicago: University of Chicago Press.

Cooley, Mike. 1987. *Architect of Bee? The Human Price of Technology.* London: Hogarth Press.

Crane, Diana. 1992. "High Culture versus Popular Culture Revisited." In *Cultivating Differences,* edited by Michele Lamont and Marcel Fournier. Chicago: University of Chicago Press.

Czitrom, Daniel J. 1982. *Media and the American Mind.* Chapel Hill: University of North Carolina Press.

Davis, Fred. 1973. "The Martian and the Convert: Ontological Polarities in Social Research." *Urban Life and Culture* 2(3): 333–343.

Douglas, Mary. 1970. *Natural Symbols.* London: Barrie & Rockliff.

Downey, Gary. 1992a. "CAD/CAM Saves the Nation?" *Knowledge and Society* 9: 143–168.

Downey, Gary. 1992b. "Human Agency in CAD/CAM Technology." *Anthropology Today* 8: 2–6.

Edgerton, Samuel. 1980. "The Renaissance Artist as Quantifier." In *The Perception of Pictures,* vol. 1, edited by Margaret Hagen. New York: Academic Press.

Ferguson, Eugene. 1977. "The Mind's Eye: Nonverbal Thought in Technology." *Science* 197: 827.

Ferguson, Eugene. 1985. "La Fondation des machines modernes: des dessins." Special issue of *Culture Technique,* 207–213.

Ferguson, Eugene. 1992. *Engineering and the Mind's Eye.* Cambridge: MIT Press.

Fleck, Ludwik. 1935. *Genesis and Development of a Scientific Fact.* Chicago: University of Chicago Press.

Fox Keller, Evelyn. 1985. *Reflections on Gender and Science*. New Haven, CT: Yale University Press.

Freidson, Eliot. 1986. *Professional Powers*. Chicago: University of Chicago Press.

Fujimura, Joan. 1988. "The Molecular Biology Bandwagon in Cancer Research: Where Social Worlds Meet." *Social Problems* 35: 261–283.

Fyfe, Gordon and John Law, eds. 1988. *Picturing Power: Visual Depictions and Social Relations*. London: Routledge.

Gans, Herbert J. 1974. *Popular Culture and High Culture*. New York: Basic Books.

Gans, Herbert J. 1985. "American Popular Culture and High Culture in a Changing Class Structure." *Annual of American Cultural Studies* 10: 17–37.

Gardner, Howard. 1984. *Frames of Mind: The Theory of Multiple Intelligences*. New York: Basic Books.

Gerth, H. H., and C. Wright Mills, eds. 1946. *From Max Weber: Essays in Sociology*. New York: Oxford University Press.

Glaser, Barney, and Anselm Strauss. 1967. *The Discovery of Grounded Theory: Strategies for Qualitative Research*. Chicago: Aldine.

Goodenough, Ward H. 1957. "Cultural Anthropology and Linguistics." *Georgetown University Monograph Series on Language and Linguistics* 9: 167–173.

Goodenough, Ward H. 1971. "Culture, Language, and Society." Reading, MA: Addison-Wesley.

Gooding, David. 1990. "Mapping Experiment as a Learning Process." *Science, Technology, and Human Values* 15: 165–201.

Gooding, David. 1993. "Visualizations in and of Science." Paper presented at the Princeton Workshop on Cognitive Approaches to Visualization in Science and Technology, Princeton, NJ, February.

Gorman, Michael, and Bernard Carlson. 1990. "Interpreting Invention as a Cognitive Process." *Science, Technology, and Human Values* 15: 131–164.

Gorman, Michael, and Bernard Carlson. 1993. "Invention as Re-representation: The Role of Sketches in the Development of the Telephone." Paper presented at the Princeton Workshop on Cognitive Approaches to Visualization in Science and Technology, Princeton, NJ, February.

Greenfield, Patricia. 1984. *Mind and Media: The Effects of Television, Video Games, and Computers*. Cambridge: Harvard University Press.

Hacker, Sally. 1990. *Doing It the Hard Way: Investigations of Gender and Technology*. Boston: Unwin Hyman.

Halle, David. 1992. "The Audience for Abstract Art: Class, Culture and Power." In *Cultivating Differences,* edited by Michele Lamont and Marcel Fournier. Chicago: University of Chicago Press.

Halle, David. 1994. *Inside Culture: Class, Culture, and Everyday Life in Modern America.* Chicago: University of Chicago Press.

Harper, Douglas. 1987. *Working Knowledge: Skill and Community in a Small Shop.* Chicago: University of Chicago Press.

Harper, R. R., J. A. Hughes, and D. Z. Shapiro. 1991. "Working in Harmony: An Examination of Computer Technology in Air Traffic Control." In *Studies in Computer Supported Cooperative Work: Theory Practice and Design,* edited by J. M. Bowers and S. D. Benford. Amsterdam: Elsevier.

Harper, R. R., and A. Sellen. n.d. "Paper-Supported Collaborative Work." Unpublished manuscript.

Henderson, Kathryn. 1991a. "Flexible Sketches and Inflexible Data Bases." *Science, Technology, and Human Values* 16: 448–473.

Henderson, Kathryn. 1991b. "Introduction: Social Studies of Technical Work at the Crossroad." *Science, Technology, and Human Values* 16: 131–139.

Henderson, Kathryn. 1995a. "The Political Career of a Prototype: Visual Representation in Design Engineering Consensus and Conflict." *Social Problems* 42: 274–299.

Henderson, Kathryn. 1995b. "The Visual Culture of Engineers." In *Cultures of Computing,* edited by Susan Leigh Star. Oxford: Blackwell.

Hindle, Brooke. 1981. *Emulation and Invention.* New York: New York University Press.

Horowitz, I. L., ed. 1964. *The New Sociology.* New York: Oxford University Press.

Hutchins, Edwin. 1991. "The Social Organization of Distributed Cognition." In *Perspectives on Socially Shared Cognition,* edited by Lauren Resnick and John Levine. Washington, DC: APA Press.

Hutchins, Edwin. 1995. *Cognition in the Wild.* Cambridge: MIT Press.

Ivins, W. M. 1953. *Prints and Visual Communication.* Cambridge: Harvard University Press.

Janson, H. W. 1977. *History of Art.* Englewood Cliffs, NJ: Prentice-Hall.

Jay, Martin. 1988a. "The Rise of Hermeneutics and the Crisis of Ocularcentrism." *Poetics Today* 9: 2.

Jay, Martin. 1988b. "Scopic Regimes of Modernity." In *Visions in Visuality,* Dia Art Foundation Discussions in Contemporary Culture, no. 2, edited by Hal Foster. Seattle: Bay Press.

Johnson, Terence J. 1967. *Professions and Power.* London: Macmillan.

Kidder, Tracy. 1981. *Soul of a New Machine.* Boston: Little, Brown.

Knorr Cetina, Karin. 1981. *The Manufacture of Knowledge: An Essay on the Constructivist and Contextual Nature of Science.* Oxford: Pergamon Press.

Knorr Cetina, Karin. 1990. "Image Dissection in Natural Scientific Inquiry." *Science, Technology, and Human Values* 15: 259–283.

Kosslyn, Stephen M. 1990. *Mental Imagery.* In *Visual Cognition and Action: An Invitation to Cognitive Science,* vol. 2, edited by Daniel N. Osherson, Stephen M. Kosslyn, and John M. Hollerbach. Cambridge: MIT Press.

Kuhn, Thomas. 1962. *The Structure of Scientific Revolutions.* Chicago: University of Chicago Press.

Lamont, Michele. 1992. *Money, Morals, and Manners: The Culture of the French and the American Upper Middle Class.* Chicago: University of Chicago Press.

Larson, Magali S. 1977. *The Rise of Professionalism.* Berkeley: University of California Press.

Latour, Bruno. 1986. "Visualization and Cognition: Thinking with Eyes and Hands." *Knowledge and Society: Studies in the Sociology of Culture Past and Present* 6: 1–40.

Latour, Bruno. 1987. *Science in Action: How to Follow Scientists and Engineers through Society.* Cambridge: Harvard University Press.

Latour, Bruno. 1988a. "Mixing Humans and Nonhumans Together: The Sociology of a Door-Closer." *Social Problems* 35 (3): 298–310.

Latour, Bruno. 1988b. *The Pasteurization of France.* Cambridge: Harvard University Press.

Latour, Bruno. 1988c. "A Relativistic Account of Einstein's Relativity." *Studies of Science* 18: 3–44.

Latour, Bruno. 1990. "Post Modern? No, Simply Amodern! Steps Towards an Anthropology of Science." *Studies in History and Philosophy of Science* 21(8): 145–171.

Latour, Bruno. 1992. "Where Are the Missing Masses? The Sociology of a Few Mundane Artifacts." In *Shaping Technology/Building Society,* edited by Wiebe Bijker and John Law. Cambridge: MIT Press.

Latour, Bruno, and Steve Woolgar. 1979. *Laboratory Life: The Social Construction of Scientific Facts.* Beverly Hills: Sage.

Lave, Jean. 1988. *Cognition in Practice: Mind, Mathematics, and Culture in Everyday Life.* Cambridge: Cambridge University Press.

Law, John. 1987a. "On the Social Explanation of Technical Change: The Case of the Portuguese Maritime Expansion." *Technology and Culture* 28: 227–252.

Law, John. 1987b. "Technology and Heterogeneous Engineering: The Case of Portuguese Expansion." In *The Social Construction of Technological Systems*, edited by Wiebe Bijker, Thomas P. Hughes, and Trevor Pinch. Cambridge: MIT Press.

Law, John. 1992. "The Olympus 320 Engine: A Case Study in Design, Development, and Organizational Control." *Technology and Culture* 33: 409–440.

Law, John, and Michel Callon. 1988. "Engineering and Sociology in a Military Aircraft Project: A Network Analysis of Technological Change." *Social Problems* 35: 384–297.

Law, John, and Michael Lynch. 1988. "Lists, Field Guides, and the Descriptive Organization of Seeing: Birdwatching as an Exemplary Observational Activity." *Human Studies* 11: 271.

Layton, Edwin T. 1974. "Technology as Knowledge." *Technology and Culture.* 15: 31–41.

Layton, Edwin T. 1986. *The Revolt of the Engineers: Social Responsibility and the American Engineering Profession.* Baltimore: Johns Hopkins University Press.

Lepowsky, Maria. 1983. "Sudest Island and the Louisade Archipelago in Massim Exchange." In *The Kula: New Perspectives on Massim Exchange,* edited by Jerry Leach and Edmund Leach. Cambridge, UK: Cambridge University Press.

Loftus, Belinda. 1988. "Northern Ireland 1968–1988: Enter an Art Historian in Search of a Useful Theory." In *Picturing Power, Visual Depictions and Social Relations,* edited by Gordon Fyfe and John Law. London: Routledge.

Luff, P., C. Heath, and D. Greatbatch. 1992. "Tasks-in-Interaction: Paper and Screen Based Documentation in Collaborative Activity." *Proceedings: Computer-Supported Cooperative Work '92* (Toronto, Oct. 31–Nov. 4, 1992), 164–170, New York: ACM Press.

Lynch, Michael. 1985a. *Art and Artifact in Laboratory Science: A Study of Shop Work and Shop Talk in a Research Laboratory.* London: Routledge & Kegan Paul.

Lynch, Michael. 1985b. "Discipline and the Material Form of Images: Analysis of Scientific Visibility." *Social Studies of Science* 15: 37–66.

Lynch, Michael. 1988. "The Externalized Retina: Selection and Mathematization in the Visual Documentation of Objects in the Life Sciences." *Human Studies* 11: 201–234.

Lynch, Michael, and Steve Woolgar. 1988a. "Introduction: Sociological Orientations to Representational Practice in Science." *Human Studies* 11: 99–116.

Lynch, Michael, and Steve Woolgar, eds. 1988b. Special issue on Representation in Scientific Practice. *Human Studies* 11 (2–3). Reprinted as *Representations in Scientific Practice* (Cambridge: MIT Press, 1990).

MacKenzie, Donald. 1990. *Inventing Accuracy: A Historical Sociology of Nuclear Missile Guidance.* Cambridge: MIT Press.

MacKenzie, Donald. 1996. *Knowing Machines.* Cambridge: MIT Press.

MacKenzie, Donald, and Graham Spinardi. 1988. "The Shaping of Nuclear Weapon System Technology," pts. 1–2. *Social Studies of Science* 18: 419–463, 581–624.

MacKenzie, Donald, and Judy Wajcman, eds. 1985. *The Social Shaping of Technology.* Milton Keynes: Open University Press.

Majchrzak, A., and H. Salzman. 1988. "Social and Organizational Dimensions of Computer-Aided Design." *IEEE Transactions on Engineering Management* 36(3): 174–180.

Manske, F., and H. Wolf. 1988. "Design Work in Change: Social Conditions and Results of CAD Use in Mechanical Engineering." *IEEE Transactions on Engineering Management* 36(4): 282–293.

Mead, George H. 1934. *Mind, Self and Society.* Chicago: University of Chicago Press.

Meiksins, Peter. 1988. " 'The Revolt of the Engineers' Reconsidered." *Technology and Culture* 29(2): 219–246.

Merton, Robert K. 1942. "Science and Technology in a Democratic Order." *Journal of Legal and Political Sociology I,* 15–26.

Merton, Robert K. 1973. *The Sociology of Science: Theoretical and Empirical Investigations.* Chicago: University of Chicago Press.

Meyers, G. 1988. "Every Picture Tells a Story: Illustrations in E. O. Wilson's *Sociobiology.*" *Human Studies* 11: 235.

Mukerji, Chandra. 1984. "Visual Language in Science and the Exercise of Power: The Case of Cartography in Early Modern Europe." *Studies in Visual Anthropology* 10: 30–45.

Mukerji, Chandra. 1989. *A Fragile Power: Scientists and the State.* Princeton: Princeton University Press.

Noble, David. 1977. *America by Design.* New York: Oxford University Press.

Noble, David. 1984. *Forces of Production.* New York: Alfred A. Knopf.

North, J. D. 1923. "The Case for Metal Construction." *Journal of the Royal Aeronautics Society* 37: 3–25.

Panofsky, Erwin. 1945. *Albrecht Dürer,* 2 vols. Princeton: Princeton University Press.

Pinch, Trevor. 1985. "Towards an Analysis of Scientific Observation: The Externality and Evidential Significance of Observation Reports in Physics." *Social Studies of Science* 15: 1–36.

Pinch, Trevor, and Harry Collins. 1993. "Inside Knowledge: The Phenomenonology of Surgical Skill." Paper presented at the Princeton Workshop on Cognitive Approaches to Visualization in Science and Technology, Princeton, NJ, February.

Polanyi, Michael. 1958. *Personal Knowledge: Towards a Post-Critical Philosophy.* London: Routledge & Kegan Paul.

Polanyi, Michael. 1967. *The Tacit Dimension.* London: Routledge & Kegan Paul.

Pye, David. 1964. *The Nature of Design.* New York: Reinhold.

Pylyshyn, Z. W. 1981. The Imagery Debate: Analogue Media versus Tacit Knowledge. *Psychological Review* 87, 16–45.

Riel, Margaret. 1989. "The Impact of Computers in Classrooms." *Journal of Research on Computers in Education* 22: 180–190.

Rolt, L. T. C. 1963. *Thomas Newcomen: The Prehistory of the Steam Engine.* London: David & Charles Dawlish.

Rudwick, Martin. 1976. "The Emergence of a Visual Language for Geological Science, 1760–1840." *History of Science* 14: 148–195.

Rudwick, Martin. 1992. *Scenes from Deep Time.* Chicago: University of Chicago Press..

Ruse, Michael, and Peter Taylor, eds. 1991. Special Issue on Pictorial Representation in Biology. *Biology and Philosophy* 6: 125–294.

Sacks, Oliver. 1993. "To See and Not to See." *New Yorker* (May), 59–73.

Salomon, Gavriel. 1979. *Interaction of Media, Cognition, and Learning.* San Francisco: Jossey-Bass.

Salomon, Gavriel, and A. A. Cohen. 1977. "Television Formats, Mastery of Mental Skills and the Acquisition of Knowledge." *Journal of Educational Psychology* 69: 612–619.

Salzman, H. 1988. "Computer-Aided Design: Limitations in Automating Design and Drafting." *IEEE Transactions on Engineering Management* 36 (4): 252–262.

Schutz, Alfred. 1970. *On Phenomenology and Social Relations.* Chicago: University of Chicago Press.

Scribner, Sylvia, and Michael Cole. 1981. *The Psychology of Literacy.* Cambridge: Harvard University Press.

Shaiken, Harley. 1985. *Work Transformed: Automation and Labor in the Computer Age.* New York: Holt, Rinehart and Winston.

Shapin, Steven. 1984. "Pump and Circumstance: Robert Boyle's Literary Technology." *Social Studies of Science* 14: 481–495.

Shapiro, D. Z., Hughes, J. A., Randall, D., and Harper, R. 1991. Visual Re-representation of Database Information: The Flight Data Strip In Air Traffic Control, *Proceedings, 10th Interdisciplinary Workshop on 'Informatics and Psychology': Cognitive Aspects of Visual Language and Visual Interfaces* (Scharding, Austria, May), Amsterdam, N.Y.

Shepard, R. N. and L. A. Cooper. 1982. *Mental Images and Their Transformations.* Cambridge, MA: MIT Press.

Simon, Herbert A. 1945. *Administrative Behavior.* London: Collier Macmillan.

Sinclair, M. A., C. E. Siemieniuch, and P. A. John. 1988. "A User-Centered Approach to Define High-Level Requirements for Next Generation CAD Systems for Mechanical Engineering." *IEEE Transactions on Engineering Management* 36(4): 262–271.

Star, Susan Leigh. 1983. "Simplification in Scientific Work: An Example from Neuroscience Research." *Social Studies of Science* 13: 205–228.

Star, Susan Leigh. 1985. "Scientific Work and Uncertainty" *Social Studies of Science* 15: 391–427.

Star, Susan Leigh. 1988. "Introduction: The Sociology of Science and Technology." *Social Problems* 35(3): 197–205.

Star, Susan Leigh. 1989. "The Structure of Ill-Structured Solutions: Heterogeneous Problem-Solving, Boundary Objects and Distributed Artificial Intelligence." In *Distributed Artificial Intelligence 2,* edited by Michael N. Huhns and Gasser. Menlo Park, CA: Morgan Kaufman.

Star, Susan Leigh. 1995. *Culture of Computing.* Oxford: Blackwell.

Star, Susan L., and James R. Griesemer. 1989. "Institutional Ecology, 'Translations,' and Coherence: Amateurs and Professionals in Berkeley's Museum of Vertebrate Zoology, 1907–1939." *Social Studies of Science* (August): 420–487.

Strauss, Anselm, and Juliet Corbin. 1994. "Gounded Theory Methodology: An Overview." In *Handbook of Qualitative Research,* edited by Norman K. Denzin and Yvonna S. Lincoln. Thousand Oaks, CA: Sage.

Suchman, Lucy. 1987. *Plans and Situated Actions: The Problem of Human-Machine Communication.* Cambridge, UK: Cambridge University Press.

Suchman, Lucy. 1988. "Representing Practice in Cognitive Science." *Human Studies* 11: 305.

Sudnow, David. 1978. *Ways of the Hand: The Organization of Improvised Conduct.* Cambridge: Harvard University Press.

Tang, J., and L. Leifer. 1988. "A Framework for Understanding the Workspace Activity of Design Teams." In *Proceedings of the Conference on Computer-Supported Cooperative Work, September 1988.* Portland, Oregon. New York: ACM.

Thune, Carl E. 1983. "Kula Traders and Lineage Members: The Structure of Village and Kula Exchange on Normanby Island." In *The Kula: New Perspectives on Massim Exchange,* edited by Jerry Leach and Edmund Leach. Cambridge, UK: Cambridge University Press.

Traweek, Sharon. 1988. *Beamtimes and Lifetimes: The World of High Energy Physicists.* Cambridge: Harvard University Press.

Turnbull, David. 1993. "The Ad Hoc Collective Work of Building Gothic Cathedrals with Templates, String, and Geometry." *Science, Technology, and Human Values* 18(3): 315–340.

Ullman, D., L. Stauffer, and T. Dietterich. 1987. "Toward Expert CAD." *Computers in Mechanical Engineering* (November/December): 56–70.

Vincenti, Walter G. 1990. *What Engineers Know and How They Know It.* Baltimore: Johns Hopkins University Press.

Whalley, Peter. 1991. "Negotiating the Boundaries of Engineering: Professionals, Managers, and Manual Work." In *Research in the Sociology of Organizations,* vol. 8, 191–215. Greenwich, CT: JAI Press.

Winner, Langdon. 1980. "Do Artifacts Have Politics?" *Daedalus* 109: 121–136.

Woolgar, Steve. 1985. "Why Not a Sociology of Machines? The Case of Sociology and Artificial Intelligence." *Sociology* 19: 557–572.

Yoxen, Edward. 1987. "Seeing with Sound: A Study of the Development of Medical Images." In *The Social Construction of Technological Systems,* edited by Wiebe Bijker, Thomas P. Hughes, and Trevor Pinch. Cambridge: MIT Press.

Zilsel, E. 1942. "The Sociological Roots of Science." *American Journal of Sociology* 47: 245–279.

Zolberg, Vera. 1984. "American Art Museum: Sanctuary or Free-for-All?" *Social Forces* 62: 376–392.

Zolberg, Vera. 1992. "Barrier or Leveler? The Case of the Art Museum." In *Cultivating Differences: Symbolic Boundaries and the Making of Inequality,* edited by Michele Lamont and Marcel Fournier. Chicago: University of Chicago Press.

Zuboff, Shoshana. 1988. *In the Age of the Smart Machine.* New York: Basic Books.

Zussman, Robert. 1985. *Mechanics of the Middle Class: Work and Politics among American Engineers.* Berkeley: University of California Press.

Index